CITY PLANNING

PERSPECTIVES IN URBAN GEOGRAPHY
Volume Fourteen

Concept's International Series in Geography

No. 1 Perspectives in Agricultural Geography
(In Five Volumes) Edited by Noor Mohammad

No. 2 Perspectives in Geomorphology (In Four Volumes)
Edited by S.H. Sharma

No. 3 Perspectives in Urban Geography
Edited by C.S. Yadav

1. New Directions in Urban Geography
2. Comparative Urban Research
3. Comparative Urbanization: City Growth and Change
4A. Models in Urban Geography (Theoretical)
4B. Models in Urban Geography (Mathematical)
5. Urban Research Methods: Central Place, Hierarchical and City Size Models
6. Contemporary City Ecology
7. Slums, Urban Decline and Revitalization
8. Contemporary Urban (Issues)
9. Rural-Urban Fringe
10. Morphology of Towns
11. Cities and Housing
12. Perceptual and Cognative Image of the City
13. Urban Economics
14. City Planning: Problems and Prospects
15. City Planning: Administration and Participation
16A. Urban Planning and Policies: Reorientation of Policy Norms
16B. Urban Planning and Policies: Cross-Cultural Urban Policies

Concept's International Series in Geography No. 3

Perspectives in Urban Geography

Volume Fourteen

CITY PLANNING
Problems and Prospects

Edited by
C.S. YADAV

CONCEPT PUBLISHING COMPANY, NEW DELHI-110 059

First Published 1987
Reprinted 2007

Published and Printed by
Ashok Kumar Mittal
Concept Publishing Company
A/15-16, Commercial Block, Mohan Garden
New Delhi-110059 (INDIA)
Phones : 25351460, 25351794 *Fax* : 091-11-25357103
Email : publishing@conceptpub.com

PREFACE

NATURE OF URBAN GEOGRAPHY

THERE are several studies on urban geography which reveal that, as a major sub-discipline, it has a long tail but a short body. However, the 'spectrum' is so vast and broad that researchers are able to make general statements in defence of its conception, philosophy, nature and orientation. Urban geography today encompasses and interfaces with various disciplines which are interested in urban studies. As an introduction to this series, our task here is to make an attempt to briefly review the development of geographical interest in various aspects of cities.

There are several reviews of the early development of urban geography which have been dealt with by Berry and Harton (1970)[1] and Carter (1974).[2] Berry and Harton in their book, *Geographical Perspective on Urban System*, have made an attempt to introduce readers to its present-day status. They have clearly resolved that "the formative years of the social science in the late nineteenth century and early twentieth century were also the years in which urban studies first developed,"[3] thus providing the context for the geographer's emerging interest in cities. However, the emergence of urban studies dates back to the writing of Greek scholars, but as a sub-discipline, it has reached it present-day status only in the past 30 years.

In their historical perspective the works on urban geography show that the pre-20th century studies primarily concerned themselves with themes of location, size and shape of the cities. The initial findings were strongly subjective, descriptive and dependent more on observation such as the works of Hassert (1907)[4] and Blanchard (1911).[5] In the succeeding years the conceptual framework of site and situation was criticized by Aurousseau (1924)[6] and Crowe (1938)[7], their conception being that cities were not inanimate objects in landscape, but also organic elements which involved people and their movements. The morphologist, later in 1960, truly brought the indigenous line of evolution in the sphere of urban geography and studies on the build-up fabric of cities (Conzone, 1960).[8] Smails (1955)[9] constituted the prime base of urban geography, which remained articulated without any major conceptual change till early 1960s. The studies on the morphological aspects of the urban system were influenced by external forces, consequently the methodological frameworks got impetus within the discipline. At this juncture the evolution of concepts was not based on environment but took shelter under the umbrella of economics of location, and incorporated analysis of land values and rents, and the concept of nodality and accessibility. These concepts were derived from the economic theories of Cooley (1894)[10], Weber (1899),[11] and Hurd (1903).[12]

The Chicago School of Urban Ecology hastened the evolution of urban geography. In his monumental work Park (1925)[13] developed the idea of order and analysis of towns. Further, a powerful thrust and much of the rationale was provided to the studies of urban geography by the Central Place Theory of Christallar (1933).[14]

The impact of the statistical method was powerful and it brought rapid and enormous changes in the field of urban geography. It also brought new insights into the development and application of urban geography towards increased quantification. Brian J. Berry, pioneer in the field, analysed the spatial order, size and location of towns and cities. There was widespread use innovative techniques to explore the nature of urban problems, hypotheses were tested, new theories propounded and old theories remodelled. The statistical methods were put to a variety of uses. Smith (1965)[13] evaluated the classification of settlements ; Berry and Garrison (1956)[16] examined the utility of the rank-size rule for urban populations. The models of Park and Burgess (1925),[17] Harris and Ullman (1954)[18] and Homar Hoyt (1939)[19] were tested and re-examined by various geographers.

A new impetus to urban geography came from social area analysis which was initially propounded by Shevky and Bell (1955)[20] and later on this provided a basis for factorial ecology.

The collaboration of this stream in urban geography was offered by Berry (1971),[21] Herbert (1972)[22] and Johnston (1974).[23] As a consequence of these thrusts in geography after 1960, the techniques of investigation were sharpened and this has provided the basis for a scientific explanation of cities. Attempts were made to introduce new theories and frame laws to make the explanation of events more rational and logical. Sophisticated models were propounded as urban geography entered a new era of rationalizing the subject matter of urban studies on the basis of new philosophies, new concepts, new theories, new methodologies and applications.

However, the status of contemporary urban geography has been elevated only recently by the behavioural approach. The studies of perception and cognition which have a long tradition in physiology were first introduced into the field of geography by Lynch (1960),[24] Dowson and Stea (1973)[25] Gould and White (1974).[26] At present there is a sudden spurt in the studies on the subject with a new paradigm. But the full impact of behavioural approaches upon urbon geography has yet to be realized.

The aim of this series is to seek reorientation of the discipline strengthened by new philosophies, methodologies, subject matters or application. The series has been arranged in such a way that all contemporary view points are covered comprehensively. Hopefully, this series will inspire researchers to appreciate the work already done by geographers in studying cities. Geography by nature seems to be a synthesizing field of inquiry. As such we have made an endeavour to combine some of the otherwise disparate facts garnered by other disciplines in such a manner that we can gain a better understanding of the urban system. The study of urban geography is essential if we are to analyse the human consequences of the settlements in which we live. It is useful to planners, decision makers in government and corporations and also to each one of us as citizens. Finally, it gives us a perspective on what may be happening to our cities and to the nature.

To achieve the above goal only those contributions with originality and contrasting viewpoints were selected for inclusion in this series. In doing so the editor does not wish to compete with the journals in the field in which innovative research and methodological aspects are presented. But he does wish to convey, and convey with conviction, that significant researchers are being undertaken in the different branches of urban geography and in other allied disciplines. The present series is an attempt to provide a selective reappraisal and rigorous examination of the assumptions and the urge to disseminate new knowledge created by the mutual interaction. Emphasis is also placed on the conceptualizatiön and theorization of the subject matter so that general laws may emerge. A conscious effort has been made to organize the series in such a way that it reflects the philosophical approach parallel to that of the behavioural school. Finally, a vigorous attempt has been made to demonstrate throughout the series how geographers are basically involved in solving the urban problems.

REFERENCES

1. Berry and Harton, *Geogrophic Perspectives on Urban System : With Integrated Readings*, Jersey, Prentice Hall, INC, Englewood Cliffs, 1970.
2. Carter, *The Study of Urban Geography*, Edward Arnold, London. 1974.
3. Berry and Harton, *op cit*,. 1970.
4. Hassert, K., *Die Stadte Geographisch Betrachect*, Leipzig, 1907.
5. Blanchard, R., *Grenoble : Etude De Geographic Urbanine*, Paris, 1911.
6. Aurousseau, M., "Recent Contributions to Geography: A Review," *Geographical Review*, 1924, 14, p. 444.
7. Crowe, P.R., "On Progress in Geography," *Scottish Geographical Magazine*, 1938, 54, pp. 1-19.
8. Conzone, M.R.G., "The Plan Analysis of an English City Centre," In. Norkorg (Ed), *Proceedings of I.G.U. Symposium on Urban Geogrophy*, C.W.K. Gleerup, Lund, 1962.
9. Smails, A.E., "The Urban Mess of England and Wales." *Transactions and Papers*, Institute of British Geographers, 1946, 11, 85, p. 101.
10. Cooley, C.H., "The Theory of Transportation," *Publications of the American Economic Association*, 1984, 9, pp. 5-7.
11. Weber, A.F., *The Growth of Cities in the Nineteenth Century: A Study in Statistics*, New York, 1899.

12. Hurd, R., *Principles of City Land Values: New York Record and Guide*, 1903, pp. 19-21.
13. Park, R.E., "Suggestions for the Investigation of Human Behaviour in an Urban Environment", in R.E. Park and R.D. Mackenzie (eds). *The City*, University of Chicago Press, 1925.
14. Christallar, W., *Central Place in Southern Germany*, Gustar Fischer, Teno, 1933.
15. Smith, R.M.T., "Method and Purpose in Functional Town Classification," *Annals of the Association of American Geographers*, 1965, 55, pp. 539-548.
16. Garrison, W.L., "Applicability of Statistical Inference to Geographical Research," *Geographical Review*, 1956, 46, pp. 427-429.
17. Burgess, E.W., "The Growth of City." in R.E. Park, and R.D. Mackenzie (eds.), *The City*, University of Chicago Press, 1925.
18. Ullman, E.L., "The Nature of Cities," *Annals, American Academy of Political and Social Science*, 1945, 242, pp. 7-27.
19. Hoyt, H., *The Structure and Growth of Residential Neighbourhoods in American Cities*, Washington, Federal Housing Administration, 1939.
20. Shevky, E. and Bell, W., *Social Area Analysis*, Stanford University Press, Stanford, 1955.
21. Berry, B.J.L. (ed), "Comparative Factorial Ecology," *Economic Geographic (Supplement)*, 1971, p. 47.
22. Herbert, D.T., *Urban Geography: A Social Perspective*, David and Charles, Newton Abbot 1972.
23. Johnston, *Urban Residential Pattern*, Bell, London, 1971.
24. Lynch, *Image of the City*, Massachusetts, Cambridge, 1960.
25. Dowson and Stea, "Image and Environment," Chicago and London, 1973.
26. Gould and White, R., *Mental Maps*, London, 1974.

ACKNOWLEDGEMENT

We Thank Editors Gerd-Michael Hellstern, Erithjof Spreer Hellmut Wollmann, *Heft 2, Vol. II* and *Vol. III*, for their permission to use material from "Applied Urban Research" Proceedings of the European Meeting on Applied Urban Research, Essen Oct. 2-4, 1981.

CONTENTS

LIST OF CONTRIBUTORS

J. Gordon Adams : Department of Town and Regional Planning, Glasgow University, Glasgow

Dimitrios S. Dendrinos : Assosiate Professor, the University of Kansas, School of Architecture and Urban Design, Kansas.

R.S. Dorney : School of Urban and Regional Planning, University of Waterloo, Waterloo, Ontario, Canada.

Karla M. Dykes : Brooklyn Association, Inc.

Michael A. Goldberg : Professor, Faculty of Commerce, University of British Columbia, Vancouver, B.L. Canada.

Helmut Guttler : Federal Research Institute for Regional Geography and Regional Planning, Bonn.

Leo Hawkins : Royal Melbourne Institute of Technology. (RMIT), Australia.

Eric S. Higgs : Limnoterra, Waterloo, Ontario (Canada).

Eric L. Hyman : Natural Resources Management Center, Quezon City, Philippines.

Gerald H. Krausse : Department of Geography and Marine Affairs, University of Rhode Island.

Ian Masser : Associate Professor, University of Sheffield, England.

Prithvish Nag : Department of Geography, School of Oriental & African Studies, University of London, London.

M.J. Oxley : Senior Lecturer in Urban Land Economic, School of Building Surveying and Land Economy Leicester LEI 9BH England.

D.R. Phillips : Lecturer in Geography, the University of Exeter, Exeter, U.K.

Harendra Pal Singh : Lecturer in Geography, M.M.H. College, Ghaziabad, Meerut University, Meerut, Uttar Pradesh.

William Shalinsky : Associate Professor, School of Urban and Regional Planning, University of Waterloo, Waterloo, Ontario, Canada.

F. Spreer : Federal Research Institute for Regional Geography and Regional Planning (BFLR), Bonn, Germany.

K. Turke : Federal Research Institute for Regional Geography and Regional Planning (BFLR), Bonn, Germany.

C.S. Yadav : Department of Geography, Bhagat Singh College, Delhi University, New Delhi.

Leo R. Zrudlo : Prof. agrégé, Ecole d'Architecture, Université Laval, Québec, Qué. Canada.

ONE

C. S. YADAV

INTRODUCTION

GROWING unemployment and the unforeseen speed in structural changes in many industrial sectors have intensified and highlighted, once more existing and growing frictions in and between different metropolitan areas. In view of the transformations, which are shaping, the future of our cities, the conventional analytical paradigms may prove to be inadequate and normative planning principles which have been propounded quite promisingly to attain a sophisticated level in the sixties and early seventies, seem under the conditions of growing unemployment, in sanitation, congestion, poverty and increasing structural changes to become questionable (Sternliels, Hughes 1975,[1] Perry, Watkins 1977,[2] Bruun, S.D. Wheeler 1980,[3] Camroon 1980,[4] Hall, Hay 1980,[5] Hud 1980,[6] Goddard 1982,[7] Vanden Berg 1982).[8] These scholars have identified that the established instruments, strategies and concepts of governmental stimulation obviously are no longer sufficient for a growth policy of present day cities. Thus in the past two decades, increasing number of scholars have shown their concern about the nature of urban life that existed in a particular city and about the strategies and institutions which tried to alter the scene in our cities.

It is a paradox that our cities are scenes of increasing affluence as well as increasing relative poverty. In the wake of affluence there has been rising demand for more space for living, leisure and travel, it has resulted in the growth of car ownership, created pressure on scarce environmental resources and accelerated the development of urban sprawl and suburbia. On the other hand poverty has generated conflict and has given rise to a decline in standard of housing, health and education and in turn have caused urban violence. As a matter of fact adequate solutions to tackle these problems have not yet been provided. Nevertheless solutions to these problems in essence depend upon the ability of a scholar to analyse the various processes which generate problems in cities and to assess the magnitude of these problems to which action can be most appropriately directed. Thus in its widest sense, it is a question of intervention or planning. The urban planning is in a way an effort to control, to guide and accomplish the physical development of towns and cities and thereby provide for the people who are living and working in them, the best possible environment. The whole spectrum of the urban planning includes in it urban unit as a whole, inclusive of the regional setting, action, intervention, interrelation, co-ordination, efficiency health utility and beauty are among its key watch words.

In the present day context, the importance of city planning is becoming more and more and steps towards it are usually appreciated because of the phenomenal growth of cities both in multiplication of their number and in their individual increase in population, as a result of which our cities are facing innumerable problems and there is haphazard growth of them. To avoid confusion and apathy, it is very essential to systematise the whole, city environment. As a matter of fact the importance of city planning is being realized because the cost of the mistakes that result from unguided growth of cities due to unbridled human interest is being impressed upto the tax payers. It is a well recognised fact that lack of proper city plan create congestion in the city, resulted in both inconvenience and loss of time and inadequate recreational space. With the help of the conscious city planning many of the forgoing ills may be avoided or eliminated. Thus city planning is a process through which systematic and orderly development of the city is achieved. As a matter of fact it eliminate chaos and create order so that citizens achieve a number of desirable facilities, convenience is attained, time is saved, congestion is eliminated, working efficiency is increased and both commerce and industry make progress. Thus all these achievements contribute to the welfare of the citizens. In a sense city planning is seeking for a better humanity through better environment. It is foresight common sense and good business judgement applied to cities. The planners consciously try to bring order without defining that order. However according to Foley (1960)[9] there are three basic ideologies which urban planners try to adopt :

1. Town planning's main task is to reconcile competing claims for the use of limited land so as to provide a consistent, balanced and orderly arrangement of land uses.
2. Town planning's central function is to provide a good (or better) physical environment, a physical environment of such good quality is essential for the promotion of a healthy and civilised life.
3. Town planning, as part of a broader social programmes, is responsible for providing the physical bases for better urban community life.

Criticism

During the last twenty years there has been a remarkable growth of diverse interest in the study of planning processes from a variety of standpoints. This diversity of interest has brought with it many problems which can be seen in the ambiguity and confusion that exists regarding the role of urban planning, theory and practice. Some scholars feel that there is no, new, neat, pat, self contained urban planning theory which can resolve inconsistencies and contradictions in people's aspirations resources and abilities. They highlight the negative aspects of the past and present planning strategies which is being qualified as inappropriate, "ineffective", "counter productive" and "destructive". Nevertheless, recently Grabow and Heskin (1973)[10] have presented three major criticism of modern planning which are also applicable to urban planning.

(i) It is elitist, establishing a rational comprehensive advisor, quite separate from the planning client. Joint participation by both planners and plannee is not part of the game.

(ii) It is a centralized activity, since it requires monitoring and control of planning action consistent with rational planning objectives.

(iii) It is a change resistant, since all changes other than that prescribed in the plan must be controlled and eliminated.

The criticism of Grabow and Heskin (1973) is well aimed because there is a contradiction between planning assumptions and realities. So they further state that unpredictable change is uncontrollable and is considered synonymous with undesirable change. All attempts are made to suppress it. But history has taught us that significant change is always unique, unpredictable and unrepeatable : change is an open mind creative process ; in the rational comprehensive model. It is precisely the creative source of social change that are not and cannot be taken into account" (Grabow and Heskin 1973).[11]

A more serious problem arises from the misspecification of problems. Planning in its process aims towards solutions of problems. Some scholars feel that it is incorrect to define problems on the basis of methodologies available at present. It creates a dilemma because the present day technique lead to inappropriate definitions and inappropriate solutions. One of the most defective framework of urban planning procedure is that planners have not provided any systematic framework with in which to cater their ends, there is no justification why these should be their ends or to fit means to ends. In brief urban planning is partially lacking in a theoretical foundation. The interdisciplinary approaches vaguely carry out researches in their own particular discipline and there is no agreement on a common framework with in which individual contributions are to be made.

Biased planning procedures

Today the aim of planning is to help people solve their problems and realize their goals. The objective of the urban planner should be to plan for community, but on an examination of procedures, it is found that it deals only with a portion of that community. Most of the planners have provided designs for land uses, the location and design of buildings, streets, transportation facilities and open space. They have primarily devoted their attention to the natural and man made physical artifacts of the city (Gans 1969).[12] The urban planners have arranged or rearranged the physical artifacts in order to create an orderly, static, efficient and attractive community. By doing so, in essence, the planners provide only a partial solution to the whole community, they have in their attempt ignored almost entirely the people who live in that community and with out them, there would be no buildings or land uses. It is a paradox that planners have not made any effort to plan for them either as individuals or as members of groups. They have completely neglected the social structures, institutions, culture and subcultures, socio-economic classes age groups and political blocks which are the vital elements of the fabric of settlements. The planners have completely forgotten that man is a social animal who has his family, the job, and friends. The planners further did not recognise, the goals, aspirations, problems and worries of a group of people, they visualize them only as occupants of dwellings, so that the main efforts of planners were devoted towards physical planning of the environment.

Now a question is posed, why has city planning taken this direction ? The answer can be found with in a historical perspective and political organization. As a matter of fact the city planning was developed as a reform movement in late nineteenth century. The objective of the reformers was to improve the terrible living conditions of the people. However, these reformers were mostly missionaries and politicians who did not have a desire to change the economy or the social order.

Although city planning in its initial stage has been principally concerned with improving the physical environment, it has also been planning for certain people although only indirectly and implicitly (Gans 1969). The objective of the planners was so biased that they planned for themselves for their political supporters and upper middle class citizens. As a matter of fact the master plans emphasized more on down town retail area. In their conception the ideal city was good for business and for property ownership. Finally it was also noticed that planners planned for people of his own class culture for middle and upper middle class professionals. While planning for these goals, the planners were not conscious that they were planning for themselves, their supporters and for people of their own class. On the contrary they believed that they were focussing on desirable type of housing, business and industry, in doing so they thought they were planning for every body. They were mistaken to believe that as professionals they had the expertise that trained them to change the community and to change people's lives according to their dictates. As a matter of fact since they were discendants of a missionary reform, they never questioned their objectives and activities.

Indeed, these master planners were not having any requisite expertise with them, their judgement were not rational based on any empirical analysis. Gans has rightly remarked "The standards and other tools used by planners in various community facilities often reflected the vested interests and cultural values and biases of the planner himself and of his allies". Thus the planners were conspirators who intentionally allied themselves with conservative political reformers and then with businessman and property owners, in order to enrich themselves and their profession.

The lack of effects of physical planning

A number of studies in the past have reflected that physical environment does not play an effective role in people's lives as the planners have thought. In spite of the fact that people reside, work and play in buildings, their behaviour is not shaped by the buildings but by the economic, social and cultural relationships. Some eminent scholars such as Wilner (1962),[13] Berger 1960,[14] Gans 1967[15] and Willmott 1963,[16] have all supported the contention that change in physical environment has not played an effective role and they have rightly believed that primary effect on people is not created by the physical environment of community, but by the social environment. As a matter of fact, the site plan and the provision of facilities had little impact on the whole community of the city. Moreover the provision of facilities for a large majority of people is unimportant because the number of people enjoying such facilities as play grounds, libraries and community centers is always small.

Although a few may be so attracted to the facility that their life is changed by it. The majority of people use a facility only rarely.

Planning for people

Although research on the effects of physical environment is limited, the available information points that these effects are not as paramount as the planners believes than the question arises how to plan for the well-being of the people. In this respect the first thing is to overlook all traditional concepts related to physical environment and to begin with the identification of how people live, what are their requirements and what are their problems to be resolved. Thus there is a greater need of social planning which guarantee the alleviation of social problems and enhancing the well being of the people (Mohammad 1977).[17] It differs from physical planning because it focusses on the collective needs and concerns of people in relation to five areas—psychological, social, economic, political and health and safety. As a matter of fact such planning procedures try to resolve financial problems through intervention in the economy to achieve greater productivity and equality of income so as to increase average income of the inhabitants. Social welfare planning also strive to resolve the familial conflict by introducing a variety of programmes. Some family conflict may be eliminated if the man could earn more, other types would be tackled by creating more jobs for unemployed people and lastly the social welfare planning ensures to reduce illness by preposing cheaper medical facilities.

The foregoing analysis indicates that planning to resolve people's problems and to help them achieve their goals does not of course, rule out physical planning but seeks to put such planning in its proper place. As a matter of fact the social and economic to which planners ought to address themselves are located in houses, offices, community facilities, although not all instances, plan for a social environment will require physical expression. Thus it can be concluded that physical plan ought to express the goals of social environment planning; housing and other aspects of the physical environment aught to be planned in terms of what people want. As a matter of fact the social and physical elements of planning are so intricate to each other that it is difficult to divide them.

Role of planner

To the urban planners there are three distinct sequential role to be played. First they are visionaries. Then they are municipal missionaries or muscular utopians and lastly they are the bureaucratic keeper of amenity. In the decade between 1960 and 1970, the planning was considered at national policy level as an evil, not an aid, and planners as amongst the most obstructive bureaucrats. However, this disgraceful assignment to planner was uplifted by the adoption of the definition of planner used by Canadian Occupational Classification Mannual (Canadian Dictionary of Occupations 1971.[18]

The mannual identifies an urban planner as some one who :

— develops programs, plans for controlled utilization and physical facilities in rural and urban areas.
— compiles, analyses and reports data on economics, legal, political, sociological and physical factors affecting land uses.
—confers with local authorities, civic leaders, social scientists, lawyers and planning and development specialists to advise and recommend zoning arrangements for land.
—prepares and presents narrative and graphic programme plans to civic and industry authorities.
—recommends, measures affecting land use, public utilities, community facilities, housing and transportation, to control and guide community development and renewal.
—supervises and co-ordinates the work of technologists and technicians.

Thus the planner's role is to lay down norms to be adopted in future and to meet the expectations of the planners. The planners should act as advisers on the allocation of land and not as dictators. Foley (1979)[19] has rightly assigned the role to urban planners in the following manner. "There needs to be a personality change in the planner—a change from the skilled expert to the participatory collaborative, listening and questioning co-ordinator. A person who can work equally with politicians and professionals---. The planner now needs to be a manager over a wide spectrum of public sector effort".

The planners can play an effective role when they act as reformer and not as a professional who is free to impose his expertise and his values on the people whom he plans. In the words of a planner (Gans 1969)[20] "as a public official he aught rather to be their servant, helping them solve their problems and achieve their goals, except when these goals have antisocial and self destructive consequences. In this process, the planner aught to propose a variety of programmes to solve problems and achieve goals so that people have maximum choice". The prime function of the planners is thus not to determine goals, for that is the duty of the politicians and of electrorate. The planner ought to concern himself principally with determining the best programmes for achieving goals and since most goals pertain to the improvement of the social and economic environments, he ought to devote himself to programmes which will achieve these goals.

ABOUT THIS VOLUME

The objective behind this volume is clear cut : It provides reader with informative, useful, readily applicable reference book on planning, programming, designing and constructing the build up environment. This volume introduces research scholar in fields directly concerned with city planning to the scope, subject matter, challenges and opportunities of comprehensive urban planning theory. At present these field include urban and regional planning, urban and regional studies, public administration systems and civil engineering. Also very much involved are economics, low, business administration, political science, sociology, urban geography, architecture land scape architecture, geology and management science. As a matter of fact only a small part of comprehensive urban planning procedures and problems can be covered in one book. To be thorough and complete on a subject that is now so extensive would require many volumes. The editor has selected those readings for the present volume which he believes represent comprehensive planning norms applicable for present day situation and for the immediate future. The intent is to introduce the reader to the scopes and content of urban planning, to interest him in its diverse subject matter and to stimulate him to explore further.

The last few years have witnessed new paradigms of planning for development. The old paradigms are rejected. The basis of this transformation was that human happiness and sense of well being would flow more or less spontaneously from the relentless pursuit of economic growth ; that economic growth would be achieved through rapid industrialization and ultimately rapid industrialization would be accommodated in a few metropolitan centers, from where its benefits, would spread to the rest of other areas. It is then obvious that rational planning growth, industrialization and urbanization could be obtained only through co-ordinated central planning. Micheil A. Goldberg in his article "Planners and Planees : Towards a Systematic Planning Process" evaluates the role of planners and planees. The paper is concerned with assumptions of rationality, certainty and comprehensiveness that underly much of current planning theory and practice. Above all it is concerned with the development of an upon ongoing planning process that seeks to resolve inconsistencies and contradictions in people's aspirations resources and abilities. The author clearly reveals that the result of such a discussion will hopefully be a combination of the dialogue recently began by a number of critics of current planing ideologies. One outcome that will replicitly be avoided is that of a new, neat, pat, self contained planning theory. Such a theory is the antithesis of the process orientation being suggested. The author has described much of the past and present planning strategy in appropriate ineffective, counterproductive and destructive.

While the analytical capacities in planning offices and administrative officers are growing, the utility of the research results to improve metropolitan living conditions seems to be more and more questionable. The author Leo Hawkins in his articles "Town planner's future role" recognises that in the modern era planners have gone through three sequential, quite distinct role phases. Firstly they are visionaries, then they are municipal missionaries, or mascular utopians and finally they are the bureaucratic keeper of amenity. The author identifies that there are pluses and minuses to living and working in large modern cities. Probably the central policy and implementation concern for those involved in managing these cities and should be how to increase the net benefit. The author in his paper looks at one of the professions involved, urban planners, and reports a study in one place, Victoria, Australia. It seeks to show how by applying role theory urban planners can better understand what is required of them now and how to prepare for future tasks.

For a period of time in the 1960s there was some interest in this relationship, but faded in the 1970s. Now it may be on the upswing again. This increased interest is vital in the opinion of Shalinskey because positive contacts between the two must be maintained and increased so that a genuine partnership can be created which will hopefully lead to a better quality planning. The author Shalinskey in his article "Relationship Between Social and Physical Planning : A Source of Concern" earnestly suggests a healthy relationship between social and physical planning. With this desire the author in his paper explores the nature of this relationship and the problems which are caused when contact is inadequate. Paradigm differences are suggested as one source of the difficulties and various aspects of these differences are explored. Then several suggestions and alternatives for improving the quality and quantity of contacts between the two groups are made in the hope that conscious efforts to reduce paradigm differences will result in closer co-operation.

During the last ten years, there has been a remarkable growth of interest in the study of planning processes of all kinds from a variety of stand points. Iam Masser recognises that this is evident in the great diversity of empirical studies which range from overall descriptions of planning processes to more detailed examination of constituent elements such as plan formulation and policy implementation. In the course of these studies a variety of subject matter has been investigated which includes both narrowly focussed local planning, project and broadly based schemes of national legislation. These studies also reflect very different academic and professional stand points which range from a primary concern with theoretical models of political power and social structure to more limited technical investigations leading to recommendations for improving the efficiency of organizational performance. The author Ian Masser in his article "The Analysis of Planning Processes : A Framework for Comparative Research" reveals that this diversity of interest has brought with it many problems which can be seen in the embiguity and confusion that exist regarding the terminology that is used in these studies. Given the assumption that planning process can be viewed on the collective response of the participants involved to a perceived planning situation, a scheme for classifying planning situations is set out by Ian Masser. In his article he assesses its usefulness in relation to the findings of a number of case studies of planning processes. As a result of this the author provides a revised scheme which sets out the main criteria that need to be considered in the selection of case study material. To work of Ian Masser may be regarded as a first step to bring together researchers from a limited number of countries who are interest in developing a consistent framework for their work on urban and regional planning processes.

A recent development in Mathematics, the theory of structural stability is presented which may influence our understanding and knowledge of phenomena and process which apply to concepts, methods and roles associated with planning theory and practice. The author Dimitrios S. Dendrinos in his article "The Theory of Structural Stability : Towards a New Paradigm for Planning" discusses this new theory, as it might contribute progressively, as a technique in planning and as a method of planning. The author in this paper ignited a new outburst of work on a topic which was initially developed in the sixties. He referred to it as Catastrophe Theory (C.T.) the theory of structure of stability. The author assumes that the phenomena of rupture, thresholds, cycles, collapses, delays, accidents, change of forms and other discrepencies are abundant in nature. Such events that are termed as Catastrophic are not only related as natural phenomena but also as social events at the micro and macro scale. Families breakup, social upheavals and evolutions occur, bureaucracies change their form. Plans go through successions of phases and form. Land use diversities and other urban forms variable exhibit is continuities over space and time. Thus the author indicates that it has become almost a triviality to conclude, as many C.T. proponents, have already done. The author suggests that to begin with C.T. is a dynamic theory. It assumes that the behaviour of a system at any point in time is under a governing optimizing criteria resulting in the system either being at rest (equillibrium state) or at dis-equillibrium. The system may exhibit in the stable behaviour or unstable one. According to the author C.T. is a branch of Topology which has never been the planners choice of sport ; pure mathematical analysis is normally discounted by planners and policy makers as too rigorous, artificially complex and lacking realism, nonetheless it is the intent of the author to provide the readers with a general exposure and understanding as to the potential and usefulness of this new theory. According to the author the application of C.T. to urban problems may influence our understanding and knowledge of phenomena and processes which apply to concepts, methods and roles and roles associated with planning theory and practice. This new theory is discussed by the author as it might contribute, progressively as a technique in planning and as a method of planning. Basic concepts of this theory are presented as the elementary constructs of a potentially new planning theory paradigm.

Within any process, contradictions can be more easily recognised if not corrected, when there is some common force linking the composite elements. Logic would suggest unity of purpose regarding the goals or net outcome of the forces. However, when the process is town planning it is the very question of purpose which tears forces apart. An interested outsider could gain nothing but dissuasive confusion by looking to planners for a definition of their goals, and quite possibly anger at realizing the non chalance with which public actions, without ordered public purpose, are undertaken. The author M.J. Oxley in his article "Economic Theory and Urban Planning" stresses the lack of theoretical basis for urban planning. The author suggests that welfare economics could provide the means for analysing the problems and a framework for developing alternative courses of action. According to the belief of the author this goal is achieved by examining the concept of welfare optimisation, the optimality of perfect competition and the nature of market failure.

A Rational Planning process adapted to address multiple objectives provides an excellent conceptual framework for environmental quality analysis. A major obstacle to this integration is the overwhelming amount of information on aspects of environmental quality. The author Eric. L. Hyman in his article "The Multiple Objective Conceptual Framework For Environmental Quality Analysis" shows the urgency that some method must be employed to collapse the information so that it can be utilized in policy making. The author feels that decision makers need summary information on the identification, prediction and evaluation of environmental quality. The author strongly supports the idea that environmental quality indicators and methods are useful in making resource allocation and regulatory decisions, setting priorities and studying changes in the state of the environment. The author indicates that rational environmental decision making must examine the full range of impacts of actions on social welfare. The social welfare function may be divided into a number of partly dependent objectives : economic efficiency, equity environmental quality, health and safety and social and political quality of life.

The late 1950s and early 1960s have witnessed a series of publications that highlighted aspects of deteriorating regional and urban environmental quality in a fashion that captured the public's attention. The general support for better environmental consciousness was linked to public health and safety. This is called ecological shrinking referred not only to a greater concern for the natural environmental but also to the notion that interactions between nature, man and his technology were dynamically interrelated. In this direction the author R.S. Dorney and E.S. Higgs in their article "Urban Valley and Classification for Open Space Planning" proposed a user-oriented methodology for classifying urban valley lands for open space planning. In many urban centres, river and stream valleys represent remaining linked natural and quasi-natural space. A holistic description of natural features in terms of their structure as well as their constraints and opportunities for development and recreation is an important but often neglected steps in urban planning and impact assessment. The methodology allows for collection, description and synthesis of abiotic, biotic, and cultural and aesthetic data. The authors made it clear that while suggested for urban valley land planning, it may be applied, with some modification to non valley land open space.

During the course of the past decades, Hong Kong housing planners have expoused the concept—widely adopted elsewhere in the world of providing accommodation in newly created communities, "new towns". At the moment, there is considerable optimism that these towns will provide healthy new living environments for their inhabitants and that they will be self contained in service and employment provision. However sceptics suggest, perhaps with justification, that although housing exists are being provided quickly, service standards adopted are broadly inadequate and will result in long term in sufficient provision in a number of related fields such as health, welfare, education and community facilities. The author D.R. Philips in his article "Traditional and Modern Health Services in Newly Planned Communities" has considered the nature of planning of an essential urban service, namely medical care, in a new town setting in a territory where the co-existence of modern and traditional medical system is a complicating factor. The author indicates that some implementable guidelines are essential for the provision of services in times of rapid planned urbanization, but it is possible to call into question some of the norms adopted in the Hong Kong new towns for medical and welfare facilities. In particular this paper suggests that traditional Chinese medical facilities would be explicitly included into the planning process in order to provide a more realistic picture of the services actually used by residents. The fact that traditional medical practitioners shops are to be found in some numbers in the new towns indicates that they are the new financially viable and supported by considerable numbers of paying patients.

One hundred years ago the living conditions for most people in Glasgow have been atrocious, even if there had been improvements since early in the century. On the other hand economic foundation of the region seemed secure. Now the living conditions for most people have improved almost beyond recognition. In real sense the people of Glasgow, through their suffering, made an essential contribution to the process of skill and capital accumulation that was part of the nineteenth century Industrial Revolution. The author J. Gordon Adams in his paper "Post War Planning in the Glasgow Region" provides a historical perspective of planning measures taken into account in Glasgow. While evaluating post war physical planning J. Gordan Adams remarks that in 1946 a Clyde Valley regional plan was published and many aspects of the present physical structure of the Glasgow region stem from decision made in that plan. According to the author the Clyde Valley Plan was also produced in a period of optimism about the possibilities of 'socialist planning'. Relating to Glasgow the Clyde Valley Plan was concerned primarily with the town problems of congestion and housing and was 'physical rather than an economic' plan. The clyde valley plan proposed that Glasgow should be constrained by a green belt. From 1925 to 1980 physical and economic planning was proposed with new paradigms, which constitutes traditional British land use planning with increased emphasis on economic and social factors. With the advent of, the Scottish Development Agency (SDA), in 1976, there seems to be enormous improvement in environment and housing conditions.

The planning of settlements has generally considered climatic and cultural factors in a superficial manner and in relatively general terms. Leo R. Zrudlo in his paper "The Role of Climatic and Culture in the Planning of Small Settlements" finds that this attitude has been more flagrantly present when the planning of small settlements has been undertaken. In town or settlement planning some work has been done on the establishment of indoor comfort criteria but little or no concern has been given to outdoor comfort. Leo R. Zrudlo in his paper focusses on the climatic and cultural aspects of small settlements, as it relates to outdoor spaces created by town or settlement layouts. In this paper Leo R. Zrudlo ambitiously makes an attempt to outline a new approach to settlement planning. This approach proposes to integrate the requirements specialists, and users into the planning process in the spirit of collaboration through consultations and participations. The author feels that the specialist have dominated the design process in the past, paying lip services to climatic factors and almost completely ignoring needs. The planning approach proposed by Leo R. Zrudlo will finally give the much neglected factors of climate and culture the importance that they deserve in the planning of settlements. The approach encourages increased co-operation through the community members better understanding of the "assumptions, value synthesis and possible flows of the system arrived at and they will be more willing to co-operate in carrying out the decision taken. Thus the community would be more involved directly in the decision processes and understand the reasons for the various decisions. A town planned in this manner has a greater chance for success than those planned by the more traditional compartmentalized planning approach.

During the next two decades the first super metropolitan city of South East Asia will emerge in Western Java, joining the world's largest civilisations with a minimum population of 10 million inhabitants. This poly-nuclear system will encompass the city of Jakarta and the surrounding satellite towns. This conglomeration of urban centres dubbed JOBOTABEK (Jakarta, Boger, Tangerang and Bekasi) is perceived in response to a series of problems facing the continued expansion of metro Jakarta. Gerald H. Krausse in his paper "JOBOTABEK : Planning for Jakarta—2000" reveals that "deconcentration strategy was adopted as a possible solution of the problems and to relieve pressures on the city. Gerald H. Krausse reveals that JOBOTABEK is the first comprehensive government sponsored approach to metropolitan planning in Indonesia. The principal aim is to identify target populations and development issues, assess their needs, while financial and administrative resources are determined. The author clearly reveals that as financed jointly by the Government of Indonesia and the World Bank, the project is action-oriented. It is intended that the end product will better reflect the concerns of government and have greater probabilities of being implemented than most consultants recommendations which may not be carried out. The author realizes that the present JOBOTABEK concept should be considered an ongoing planning process atleast until the turn of the century. The author believes that it is indeed only a framework for more detailed planning by various jurisdictions and agencies responsible for providing infrastructure and services.

Dr. Prithvish Nag in his paper "Lusaka. Problems in Developing an African Capital" finds that a capital city of south Central Africa seems to be the most complicated urban centre of Zambia. The author is justified when reveals that the approaches in planning, complicated land holdings, unplanned settlements, statutory and unstatutory development, gaps between the built-up areas and social and economic segregation of different quarters have made urban development process for the city an intricate exercise. To get rid of these problems Lusaka city council has engaged Doxiadis Associates for the preparation of a development plan. According to Prithvish Nag, this capital city has a bright future in south central Africa and will continue to attract population and development. In has proved to be a growth pole for Zambia the Author identified that proper planning is essential keeping in view of its national importance as a highest order service centre.

Dr. Harender Pal Singh in his paper "Metropolitan Planning within the Framework of Regional Planning in Developing World" clearly visualizes that metropolitan planning is very much similar to regional planning as it is based on the principle of futurology. To check metropolitan growth is the beginning and end of the 'communis opinio doctorum'. According to him it is also framed and practiced in spatial, temporal and sectoral framework. The spatial perspective and metropolitan planning adopts the 'micro', the 'macro' and the 'mega' approaches to deal with urban problems in a comprehensive manner. Sectoral and temporal aspects tackle the problems in a more systematic way. In this paper Dr. Harender Pal Singh attempts to derive conclusions from the case studies from the developing countries. He is justified while making remarks that disintegration of physical and economic planning of the metropolis with the national level is the main problem which hampers the speedy solution of the urban problems. The metropolitan areas are suffering from high density of population, congestion, lack of transportation, and invironmental pollution. The respective governments are busy to solve these problems on footwar scale. But consistent migration wanes and high rates of population growth make all the estimates and projections false.

According to the author development of ring towns seems proper solutions as it checks migration.

The Federal Republic of Germany used to be a world's trade giant, always boasting of a sound surplus in external trade balance. In the last ten years the situation has changed dramatically. While cost for imported mineral oils soar from 10 billion DM in 1970 upto some 65 billion DM in 1980, the balance of payment reached a negative record of 28 billion DM last year. In its endeavour to tackle the energy problem, the Federal Government has developed an energy programme in the course of the 1970s. The authors F. Spreer and K. Tinke in their article "Local and Regional Supply Concepts : A Challenge for Urban and Regional Planning" reveal that in order to implement a strategy for energy saving macro-economic planning and national energy policies have to rely on thoroughly analysed regional and local energy-related conditions and parameters, as well as on local decision making in the public and private sectors that are in line with a sensible utilization of energy. For both sides of activities—analysis and implementation applied urban and regional research plays an important part. This research functions in two directions (*i*) in analysing local situation it can report detailed, implementation oriented information back to decision makers on the state and national levels, (*ii*) by disseminating ideas and practical level, it helps pursue national energy policy objectives.

The existence of shortages of traditional energy sources poses a design challenge to architects, engineers and social planners. Energy shortages and controls on the use of energy will have an ultimate effect on urban planning inclusive of the location of structures and the uses and functions to be served, the design of industrial, commercial and residential buildings, the transportation systems spatial uses and over all living patterns of people in urban and rural areas. The author Karla M. Dykes in his paper "The Limits to Energy-Sensitive Land Use Planning : Problems and Solutions in the United States" reveals that data on how much energy can be saved by the application of land planning techniques is incomplete, still land use tools can help to realize the full potential of energy conservation. Amassing data and its distribution will be topic of this paper. It has been shown that minor changes in layout, design and land scaping can have major conservation effects. Much of the energy consumed in the residential and transportation sectors is directly affected by individual consumer's personal habits and behaviour. The non-technological problems will prove to be the most challenging of all. Local energy programmes can be more effective than federal or state programmes by involving the community in energy utilization changes. Mechanism available to local municipalities for implementing energy conservation measures will be discussed i.e., zoning and subdivision ordinances, density and design. The author finds that designers and builders agree that the codes are the major block to energy economy, removing barriers to energy conservation will be discussed in the context of city-regional scale and neighbourhood housing unit scale. The findings it survey studies undertaken to identify and investigate specific kinds of land use control provisions which are most effective in reducing energy consumption will be presented. Two outcomes of this research are (*i*) to determine the extent and character of energy sensitive land use control implementation by local governments, (*ii*) duplication of pilot studies in other regions and shared information on findings between municipalities.

Development planning in the Federal Republic of Germany is structured in two levels (*i*) the land use plan gives a general framework of land use on the basic of urban development objectives for a medium or long term period. The land use plan is not immediately legally binding for citizens but it binds communes as well as sectoral planning agencies. The second step is the local plan. It is developed out of the land use plan and gives a detailed determination of urban land use which is binding for every one. The author Helmut Guttler in his paper "The Application of Land Policy Instruments in Urban Planning" tries, starting from a definition of land policy, to discuss the fiscal and legal instruments of land policy to improve regulate and implement land policy goals in accordance with the urban system in the Federal Republic of Germany. The following elements of land policy are discussed by the author :

— Fiscal instruments : Building Land Tax, Property Tax, Increment Tax.
— Legal Land Policy Instruments : Federal Building Law, Federal Urban Renewal Act.
— Legal Instruments in the Implimentation of Land Policy.
— Legal Instruments in the Establishment of Land Policy.

In addition land supply policy will be discussed and further development of real estate legislation and land policy instruments will be explored.

REFERENCES

1. Sternliels, G. ; Hughes, J.W., (1975) Post-Industrial America : Metropolitan Decline and Inter Regional Job Shofts, New Brunswick.
2. Perry, D.C. ; Watkins, A.J. ; (1977) "The Rise of the Sun Belt Cities". *Urban Affairs Annual Reviews.* Vol. 14 – Beverly Hills, London.
3. Bruun, S.D. ; Wheeler, J.O., (1980) The American Metropolitan System. Present and Future, Arnold.
4. Camroon, G., (ed). (1980) The Future of British Conurbation : Policies and Prescriptions for Change. London.
5. Hall, P.; Hay, D., (1980) Growth Centres in the European Urban System, London.
6. Hud, (1980) Office of the Policy Development, Changing Conditions in Large Metropolitan Areas, Washington D.C. (HUD—PDR 637).
7. Goddard, J.B. (ed) (1982), The Urban and Regional Transformation in Britain, London.
8. Vanden Berg, et al. Urban Europe : A Study of Growth and Decline, Vol. 1 of the CURB (Cost of Urban Growth) Project, Oxford, New York, 1982.
9. Foley, D.L., (1960) "British Town Planning : One Ideology or Three", *British Journal of Sociology*, 11 (3), 211-231.
10. Grabow, S. and Heskin, A., (1973) "Foundations for a Radical Concept of Planning", *Journal of the American Institute of Planners*, 39 (1) p. 108.
11. Op. cit.
12. Gans, H.J. (1969) "Planning for people, not building", *Environment and Planning*, Vol. 1, pp. 33-69.
13. Wilner, D., Walkeley, R., Pinkerton, T. and Tayback, M., (1962) Housing Environment and Family Life, John Hopkins University Press.

14. Berger, B., (1960) Working Class Suburbs, University of California Pfess.
15. Gans, H.J., (1967) The Lenittowners, Allen Lame, The Penguin Press.
16. Willmott, P., (1963) The Evolution of a Community, Routledge and Kegan Paul.
17. Mohammad, A. Qadeer, (1977) "The scope of social planning in urban planning", Plan Canada, 17, (June 1977) 86-95.
18. Canadian Dictionary of Occupations, 1971, Vol. 1, Ottawa Information Canada.
19. Foley, J., (1979) "Opinior", *Royal Australian Planning Institute Journal*, Vol. 17, No. 4.
20. Gans, (1969) *op. cit.*

TWO

MICHAEL A. GOLDBERG

PLANNEES AND PLANNERS
Toward a Synthetic Planning Process

INTRODUCTION

THIS paper is concerned with people who do planning (planners) and the people for whom planning is done (plannees). It is concerned with assumptions of rationality, certainty (or certainty equivalents), and comprehensiveness that underly much of current planning theory and practice. Above all it is concerned with the development of an open ongoing planning process that seeks to resolve inconsistencies and contradictions in peoples' aspirations, resources and abilities. The result will hopefully be a continuation of the dialogue recently begun by a number of critics of current planning ideologies. One outcome that will explicitly be avoided is that of a new, neat, pat, self-contained planning theory. Such a theory (a product in essence) is the antithesis of the process orientation being suggested.

One word comes to mind to describe much of past and present planning strategy : inappropriate. The frequent mismatch between planning methodologies and human problems has led to a variety of difficulties. At best planning often has been ineffective. At worst it has been counterproductive and destructive.

The inappropriateness of past planning concepts and values

Untenable assumptions abound in many of the social and policy sciences. The contradiction between assumptions and realities should provide a powerful base for changing assumptions, and methods, and thus reality. Too often in the past there have been attempts to change reality instead to meet untenable assumptions. This has led in some instances to a questionable reality (witness freeways and urban renewal). Accordingly criticism has become increasingly frequent and well aimed.

One of the most compelling recent criticisms of planning can be found in Grabow and Heskin (1973). They present three major criticisms of modern planning :

1. It is elitist, establishing a "rational comprehensive advisor" quite separate from the planning client. Joint participation by both planner and plannee is not part of the game.
2. It is a centralizing activity, since it requires monitoring and control of planning actions consistent with rational planning objectives.
3. It is change-resistant, since all change other than that prescribed in the plan must be controlled and eliminated.

> "Unpredictable change is uncontrollable and is considered synonymous with undesirable change. All attempts are made to suppress it. But history has taught us that significant change is always unique, unpredictable, and unrepeatable : change is an open-minded creative process ; in the rational-comprehensive model, it is precisely the creative source of social change that are not and cannot be taken into account".[1]

Another source of criticism stems from the misspecification of problems. Planning (and technology more generally) often comprises a set of solutions in search of a problem. To define problems in terms of the methodologies available to analyze them is incorrect. It leads to inappropriate definitions and inappropriate solutions. Even worse, it can lead to larger and yet more intractable dilemmas than those with which we started (Cartwright, 1973).

Post-War foreign aid and economic development schemes serve as a good example. Foreign aid and development programs have been styled on the western model, oriented toward efficiency and large-scale enterprise and rapid industrial development. Such programs have had a strong urban bias, thus aiding the flow of people into overcrowded cities.

Western aid plans have been expensive and counterproductive in a number of ways. First, they have forced traditional societies to adopt western technology and western values. Previously self-sufficient areas thus get plugged into the "world economy" and become tents in the global village. Once hooked on western technology, western products and western markets, traditional conservative production methods in agriculture, manufactures and services (e.g. medicine, religion and culture) become swamped by a need for large-scale production to supply foreign and domestic markets. Foreign markets are needed to supply foreign exchange so that "industry" can develop and foreign aid can be paid back. Domestic markets allow for import substitution which compounds the problem. Once plugged into such a Catch-22, there is little change of getting out as long as one buys the underlying assumptions of the "catch". Do developing nations *really* need to change all of their age old methods of production and consumption? If they don't, then why do they need foreign exchange, and so on and so on . . .

Redefining the problems of underdeveloped countries in their own terms is a good place to look for more appropriate approaches to planning. Perhaps the "traditional" nature of these societies is their strength, not their weakness. The high efficiency, large-scale "top-down" change advocated by western planners and decision-makers just might be as counter-productive in peasant agrarian societies as we are seeing it to be in "post-industrial" cultures.[2]

In a most comprehensive and illuminating recent paper John Friedmann and Mike Douglass (1975) summarize the aims and misfortunes of traditional top-down economic and urban-oriented development planning and its spatial adjunct, growth poles.

> "The last few years have seen a reaching out for new paradigms of planning for development. The old paradigm is no longer compelling in its logic. The underlying idea

> was that human happiness and sense of well-being would flow, more or less spontaneously, from the relentless pursuit of economic growth ; that economic growth would most readily result from rapid industrialization ; that rapid industrialization would be accommodated most efficiently in a few metropolitan centers from where its benefits would "spread" to the rest of the national territory ; that planning for growth, industrialization, and urbanization could be accomplished only through coordinated central planning ; and that coordinated central planning was a technical task that would be solved through complex econometric modelling joined to sophisticated administrative technique . . .
>
> In both its non-spatial and spatial dimensions, however, this paradigm has come under fire. On one hand, it is argued that the paradigm has failed to bring about the results expected from it. On the other hand, the last few years seem to be signalling the transition of the international economy to a new state of affairs which suggests that the unmodified pursuit of the old paradigm would be, to put it bluntly, suicidal."[3]

The doubts expressed by Friedmann and Douglass have been echoed by many others of late. In addition to the mere failure of economic development planning, growing fears are being expressed about the destruction of local peasant economies and societies which the further pursuit of narrowly defined economic goals will surely bring on. Because of the increasing realization of these fears a number of alternative approaches to development planning have been proposed.

Recent advances in planning strategies :
The continued case of development planning

Recently, thinking has begun to move away from emphasis on capital investment in cities and promotion of export markets, to a more balanced view placing increasing importance on human beings living in rural settings, producing for their own or for domestic purposes. This three-fold shift away from capital, cities and mass high-technology industrial production can be seen in a number of the suggested improvements to current development planning theory and practice. A major and influential example is provided by the "agropolitan development" idea of John Friedmann and Mike Douglass (1975).

Friedmann and Douglass refer to the post war development policies as constituting "a strategy of accelerated industrialization". Agropolitan development in contrast is termed" strategy of accelerated rural development".

To operationalize the concept, Friedmann and Douglass see at least eight major objectives. In abbreviated form these include :

1. Transforming the countryside by selectively adapting appropriate aspects of urban areas to rural ones. The hoped-for result : an agropolis or city-in-the-fields.
2. Creation of agropolitan districts which extend social, economic and political links beyond single rural villages.
3. Reduction of social dislocation by providing for development and change in rural and therefore familiar settings.
4. Stabilization of rural and urban incomes and reduction of income disparities among them.
5. Improving the utilization of human and natural resources within each agropolitan district through a more intensive development of these resources.
6. Linking of agropolitan districts through the provision of intra-structure and the regionalization of certain higher order functions.
7. Decentralization of decision-making to the local and agropolitan district level to provide for more appropriate decisions consistent with local resources, needs, desires and constraints.
8. Provision of suitable financing for agropolitan development through local retention of savings, transfer of funds from metro to agropolitan centers and improvement in the terms of trade of rural areas.

These objectives serve to emphasize the change in focus from urban to rural development. This change in emphasis is mirrored in the works of many regional planners during the past several years.[4] A particularly interesting addition to this growing literature is the one provided by Kenneth Ruddle (1975) in his view of agropolitan development from the perspective of the rural hinterland. Essentially, Ruddle is arguing from experience in favor of a bottom-up approach to regional development planning. As such, he takes the broad prescriptions of Friedmann and Douglass and others, and places them in the context of rural areas in developing countries.

More specifically, Ruddle stresses the following issues :

1. How do local and larger regional systems interface ?
2. How capable are local institutions of fostering rural development ?
3. How is knowledge transmitted in the traditional educational system ?
4. How are decisions made locally ? How are people included or excluded ?
5. How much capacity for growth does the traditional agroecosystem possess ?

Ruddle believes that to bring about meaningful change at the local micro-planning level, these questions must be addressed and answered. However, he also raises a more fundamental question, and one which provides a link with the ideas that follow. He questions who should be doing the planning in the first place :

> We must then ask ourselves who should define the change from "traditional" to "modern". Should it be done by the planner, the observing social scientist, or by those who constitute the so-called "client population" ? Recent trends in our thinking indicate that we are moving toward a combined "local-extra-local" team of definers, decision-makers, planners, scientists, and representatives of the clients. For clearly the development process is a two-way street : both developees (periphery) and developers (centre) have much to learn from each other.[5]

Locally based decisions : Constraints, virtues and hopes for the future

Orienting programs for social change toward widely dispersed rural peasant populations faces a number of social and institutional constraints.[6] It is also more difficult to administer such change within the context of the typical western administrative scheme. Such management methods are best applied in large-scale undertakings that are reasonably well defined in terms of their objectives and geographic boundaries. It is no wonder that western aid programs have been aimed at urban development where large-scale projects can be undertaken and administered using tried (and not-so) true western management methods.[7]

Constraints

A number of contraints are posed by locally based decisions. Some might turn out to be blessings in disguise.

Change must be slow : Peasantcultures have persisted in part because of their resistance to change and to innovation. Elaborate rituals exist in large measure to maintain the known *status quo* in favor of the unknown "progress" which innovation holds out. Innovations must proceed slowly and made part of the fabric of local cultures so as not to disrupt traditional methods of doing things. "Business *almost* as usual" should be the rule.[8]

Change must be helpful : Innovation cannot be carried out on a devil-may-care basis. New approaches to providing for the health and well-being of indigenous populations must carefully be screened in advance to rule out changes which have high components of risk. With limited resources, traditional societies can hardly afford to gamble on risky innovations. They will be more likely to try new ideas if the chances of achieving the hoped-for ends are reasonably high, Risk absorption might be one of the roles for future international aid.

Change must lead toward self-sufficiency : Self-sufficiency should be a high priority goal of future "development" programs.[9] Only through self-sufficiency can rural traditional populations buffer themselves against the forces of nature and the forces and whims of world markets. By striving for self-sufficiency, localized populations can achieve much of the dignity and self-respect stripped from them by central governments, generally well-meaning foreign aid, and the rigors of the market economy. If local units are self-sufficient and free to generate surpluses in commodities or services in which they might have some natural advantages, then the nations and regions of which they are a part will, in turn, move toward self-sufficiency and increased freedom, dignity and autonomy.[10]

Change must be decentralized and diverse : Elsewhere, I have sketched out the need to proceed cautiously via small and slow changes.[11] Planning through the creation of a centralized agency I suspect would be counterproductive as such an agency would quickly lapse into the convenient mode of operations for centralized organizations : big and fast. By decentralizing we can begin to meet local needs, and also begin to resist the dangers of large-scale (centralized) success.[12] More importantly, we can develop diverse solutions to problems. Such diversity is an excellent way to experiment with new ideas, which when combined with "small and slow" avoids the dangers of disruption and potential disaster which can plague larger-scale quickly implemented changes. Diverse, local, small-scale experiments are essentially the manner in which natural evolution proceeds, and harmony with nature is ultimately an important long run measure of the usefulness of change.

By encouraging (or enabling) diverse and highly localized decision rules to hold sway it is unlikely that all members of a society will make the same decision. With large numbers of different small-scale decisions we again preserve diversity, stability and persistence.[13] We might also avoid the prairie farmer fallacy of composition which has plagued market-oriented large-scale agriculture. Using similar decision rules (be they each relatively small in scale and decentralized) farmers wind up planting similar crops that have high values at the time of planting, leading to abundance at harvest time and lower prices. The effect is the same as if the decisions were taken by a centralized decision-making body on a large scale.

One other aspect of decentralized planning is that those coordinating functions that must be centralized for national planning purposes can concern themselves with broad strategic issues and ensure that centralized functions are means toward regional development ends, not ends in themselves. Central functions can provide umbrellas under which local efforts can flower.

Change must be accessible to the people : Innovations or changes in production practices, if small and slow, non-disruptive, oriented toward local needs and skills, and diverse and decentralized, can still go for nought if the people who are affected are not capable of utilizing the innovations because they require knowledge or expertise beyond their reach. Accordingly, changes must be in such a form that local populations can understand how to carry on and maintain the innovative process themselves. Change and the process of change must be communicated in a manner which is most easily understood by uninitiated people. Simple language, simple technologies and simple ideas must be the rule.[14]

Change must ultimately be peasant-based (bottom-up innovation) : The foregoing constraints are not likely to lead to a body of technology and knowledge, or to a set of institutions that resemble anything western societies and planning techniques thus far have evolved. A key ingredient in the development of a viable, non-disruptive, humane and accessible planning and innovation process lies in the answer to the question, "Who will be the initiator and champion of the process and of future change ?" If change must continually be coaxed out by central government actions it will soon die out. If, however, it is the result of locally based innovation it is likely to persevere. A "bottom-up" approach to the planning and development process is unlikely to violate the constraints delineated above, since the local residents themselves provide at one and the same time the constraints and innovations. Societal schizophrenia (or "dualistic dependencies", etc.) is a facet of western culture and not of the traditional cultures of the underdeveloped world.[15] It is unlikely that locally based, locally initiated, and locally implemented decisions and changes will violate local needs, resources, skills and natural carrying capacities. Traditional societies have after all "traditionally" been self-sufficient and persistent. A return to localized bottom-up planning, development and decision-making can blind upon these "traditions".

Virtues

By following a process of change which embodies the ideas set out above. I would hope that a number of benefits would be derived. The expected benefits would include both avoidance of problems caused by the acceptance and application of western models of development in non-western settings, as well as a variety of positive gains in the well-being of the world's rural peasant populations.

Change will be small and slow : The value of small and slow changes has been brought out in some detail in other places.[16] Small and slow changes can avoid disruptions to traditional societies. Doing minimum social, economic and environmental damage in delicately balanced traditional cultures results in minimum intrusion into the socio-ecological systems of these cultures and should provide a reasonable chance for survival and basis for innovation that is durable and helpful. Small and slow changes are also the only kind likely to be tolerated, accepted and maintained by conservative rural agrarian cultures.

Change will be diverse : The diversity of innovations which is likely to result from the above approach can have numerous beneficial consequences. Innovations will be geographically dispersed, spread through a number of economic activities, and have a variety of different cultural manifestations. Such diversity and dispersion of innovations should largely avoid the problems created by rapid urbanization of rural people fleeing the countryside. It should enable rural peasant populations to evolve traditional production techniques to the point where they can begin to achieve self-sufficiency and perhaps even surpluses for bartering or trade. Economies and social systems can develop with a balance that is impossible under the high efficiency criteria of western approaches to development. Self-sufficiency will be attainable. The simplistic single industry or single crop mentality that has brought such potential instability to western agriculture and industry will not obtain and greater stability is likely to prevail. By creating diverse and balanced settlements throughout traditional societies, the various rural settlements can become increasingly buffered from the hardships wrought by being connected to unstable world markets in primary products from agriculture to non-renewable resource-based industries.

Change will be under the control of those affected by it. The subjects of change will be the agents of change. This again would appear to be one of the principal tenets of a just and humane society. People who control change and technological and social innovation are unlikely to be imposed upon by "experts". They will be the experts. Notwithstanding the inevitable mistakes that accompany innovations, if the innovatees are the innovators they are much less likely to be the victims of innovations, since they will have it within their power to experiment further and moderate their evolving innovations so as to minimize the intrusion on their well-being and maximize the benefits to be derived from change.[17]

Change will avoid the western catch-22. Self-sufficiency, local innovation based on local needs, skills and resources all of these "bottom-up" approaches should move toward avoiding the dangers of the "top-down" philosophy of western development schemes. By orienting innovation to local cultural needs, values, etc., if nothing else rural societies can begin to become disconnected from the vicissitudes of world markets and tastes. By having rural agriculture producing foodstuffs by and for local people (with surpluses to the cities), underdeveloped countries can avoid the treadmill of "market agriculture". By producing foods in traditional ways, with innovations to improve yield and quality, and by avoiding large-scale high technology "progress", such as mechanisation in agriculture and manufactures, traditional rural economies can persist and evolve. They can be labor-intensive, and through selective local improvements in growing and harvesting techniques and in manufacture, they can survive and begin to prosper as they start to produce surpluses to sell to the urbanized areas in their countries or regions.

Labor-intensive production keeps them "back on the farm" and out of cities.[18] It cuts down the need for foreign-produced machinery, urban planning technology, fertilizers, herbicides and pesticides and cuts down the need for foreign exchange. Lower foreign exchange requirements means that there is still less pressure on production for foreign markets and more labor and resources can be freed for local or national production. There is less pressure to move to cities and less money needed to build cities. Labor-intensive production makes use of the one truly abundant resources in underdeveloped countries : people and their native skills. Labor-intensive enterprise based on traditional products and traditional techniques (as a starting point from which innovation can proceed) also can significantly ease the endemic unemployment ills of underdeveloped countries.

Above all, by building upon tried and true traditional skills and abundant labour, the need for foreign capital is drastically reduced. Reducing capital requirements goes a very long way toward getting off the treadmill. The lure of capital grants or low interest loans is illusory. Even if the funds are made available as a gift, funds are virtually never provided to maintain the stock of capital purchased with the "gift". Parts and skilled labor are needed to maintain the machines and these invariably must be purchased in the world market-place again.[19] Giving capital grants to underdeveloped countries can be linkened to "giving" free heroin to an addict. There are few gifts more dangerous than those which are addictive. The gift merely perpetuates the need for more of the same. Unfortunately, "more of the same" costs money, usually foreign, and so we are back on the treadmill of western development. Maintenance of traditional values and techniques of production can be adapted according to local needs and skills without the wholesale adoption of western values, products and technology. To avoid the sales pitch surrounding western "economic miracles" is to buy an option on a truly viable, if somewhat less materialistic, future. Western large-scale economic development is merely the Trojan Horse grown global.

At the heart of the foregoing suggestions is a call for a reorientation of our planning processes, from an external determination of what's wrong to an internal (local) determination of what's right. Instead of "problem-oriented" approaches, a more suitable foundation for change might be a

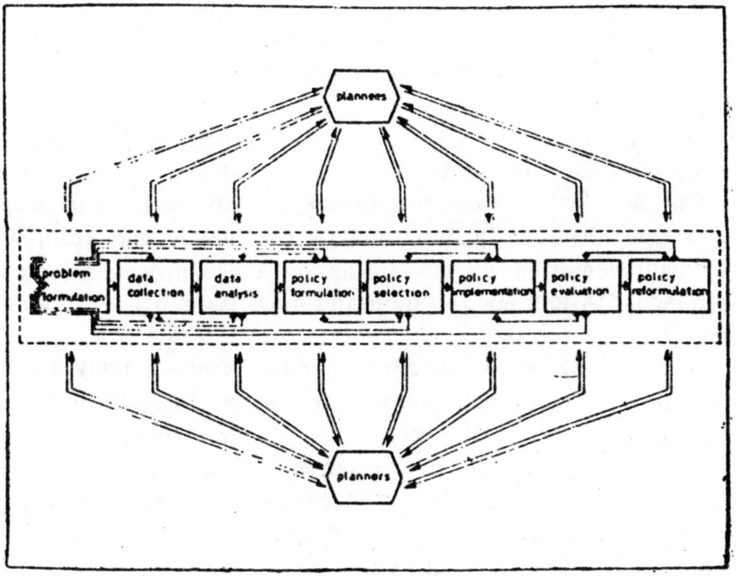

FIG. 2.1 : *Stable and unstable states*

locally specified inventory of what works well, what doesn't work so well, and what doesn't work at all and really needs changing.

Most importantly, planning should be viewed as a process ; a continuing activity, fluid and dynamic, itself capable of change. Figure 2.1 sets out an idealization of such a process. The process is ongoing, in a continual loop. Plannees and planners both have inputs to all facets of the process as well as outputs. The process is a mediator between plannees and planners. The result of continuing interaction along the lines of Figure 2.1 is not a "plan" (e.g., a product). Rather it is an approach to making decisions based on local and external perceptions and values. It is a process which, if realized, has a capacity to initiate change, guide, monitor and evaluate it, and to change itself over time as new needs are identified. Participation in this kind of planning could provide an open-ended learning experience for both plannees and planners.[20]

I. CONCLUDING OBSERVATIONS : PUTTING THE CONCEPTS TO WORK TOWARD MORE FLEXIBLE AND DURABLE DEVELOPMENT PLANNING : A SUMMARY

Ultimately, the only way to get the process of synthetic planning underway, is to do it. Each village, town, city, region, and nation must recognize a need that is not currently being met satisfactorily by "modern planning", and respond to this recognition by embarking on a new process-oriented approach. "A journey of a thousand miles starts beneath one's feet."[21]

The stock of our knowledge to date results largely from trial and error. I am arguing for a new set of efforts, directed by local populations, in concert with national efforts, to add to our stock of useful knowledge through a program of local experimentation and learning. Much can also be gained by learning what not to do. Knowing what to avoid is perhaps as important as knowing what to do.

Recent suggestions for change in planning seem to be turning past planning practice upside down. Instead of top-down planning, a bottom-up approach is being widely suggested. Instead of capital-intensive development, labor-intensive ideas are being put forward. Decentralized planning is being argued for at the expense of centralized planning, and now emphasis on rural areas is being urged instead of our former focus on urban regions. To begin to summarize the arguments in this paper I would like to suggest some other inversions of planning truisms.

1. *Development through independence, not independence through development.* As we have seen, and as others have documented,[22] the past strategy has led to increasing dependence on world markets, western technology and western sources of capital. The result has neither been independence nor development. By turning the causality around it is conceivable that past difficulties might also be reversed. Self-sufficiency is tne kind of development being argued for here.

2. *Generating trade through production of surpluses, not generating surpluses through trade.* Once again this is a reversal of past strategy and extends the previous suggestion. Self-sufficiency seems to be a strong base for trade. Moreover,

if local and regional units become reasonably self-sufficient and trade surplus production they can be buffered against the vicissitudes of external markets. An adjunct of this reorientation is a movement away from high volume specialized production for external markets, back to the more generalized production methods that characterize self-sufficiency. Specialization breeds vulnerability.

3. *Setting of priorities and production methods at the local level.* This is an essential element in achieving self-sufficiency and in determining where surpluses can reasonably be generated. It achieves a number of desirable ends. First, it allows production to be reconciled with local needs, abilities, and resources. It avoids the inconsistencies and conflicts which characterize much of centralized production planning. Avoiding such inconsistencies at the local level bodies well for the allocation of resources at more global levels, since if resource use is rationalized locally, it can also be rationalized on a more global scale.[23] A second virtue of local production autonomy is that it allows for the use, and therefore perpetuation and refinement of local skills and know-how. The movement toward self-sufficiency is done as a result of local knowledge ; not at its expense, as is the case when moving toward high technology production for external consumption.

4. *Education through experience and tradition.* Traditional educational systems were based upon experience and learning from experience. Western modes of development necessitated western educational systems based on learning abstractions and deductive reasoning. The emphasis on school-based learning was expensive in two ways : it required large sums to build and maintain ; it threatened to destroy the stock of knowledge built up over milennia. If we move back toward self-sufficiency grounded in traditional knowledge, the educational process can also be reoriented toward one combining traditional skills with experimentation based on abstraction and deduction. A synthesis of traditional and western education is thus possible, which in turn could bridge some of the gaps that are currently being created between traditional and western cultures.

FROM THE SPECIFICS OF DEVELOPMENT BACK TO THE MORE GENERAL CASE OF PLANNING

Development planning constitutes a specific, and clear-cut instances of the inappropriateness of planning as it is presently practiced. The mismatch between planners and clients in developing nations and regions is parallelled by a similar mismatch in the cities of developed nations. In each case the role of the plannee has been degraded and demeaned. Instead of a world where the customer (the citizen) is always right, we have created a world where the expert and expertise itself is always right, and the citizen and basic human needs have been largely forgotten.

Whether we talk of squatter settlements or urban homesteading, monocultural agriculture or monofunctional zoning, rural or neighbourhood development, etc., the roots of the problem are similar and lie in our latent distrust of clients and our presumptions of knowledge about the well-being of others.

> Sometimes it is said that man cannot be trusted with the government of himself. Can he, then, be trusted with the government of others ? Or have we found angels in the forms of kings to govern him ? Let his history answer this question.[24]

The same structures apply to the process of planning as to the process of government.

A series of constraints were noted above in relation to allowing local traditional cultures to take part in planning change and their own alternative futures.

These include :

1. Change must be slow.
2. Change must be helpful.
3. Change must lead toward self-sufficiency.
4. Change must be decentralized and diverse.
5. Change must be accessible to the people it affects.
6. Change must be locally based and controlled.

To balance these constraints four virtues were identified as accompanying locally based planning.

1. Change will be small and slow.
2. Change will be diverse.
3. Change will be under the control of those affected.
4. Change will avoid the catch-22 of dependence upon outside experts wherein larger technical solutions require more experts, leading to larger problems etc.

It is obvious that locally based (bottom-up) planning will have the same characteristics in the neighbourhoods of cities in developed countries as in the villages of rural areas in developing nations. The critical ingredient is the esteem accorded the client—the plannee. It has been the argument of this paper that erasing the dividing line between plannees and planners can begin the process of restoring the dignity of the client and assuring the humanity and utility of the plan and of the process.

> As long as the spell of the objective consciousness grips our society, the regime of experts can never be far off ; the community is bound to remain beholden to the high priests of the citadel who control access to reality. It is, at last, reality itself that must be participated in, must be seen, touched, breathed with the conviction that here is the ultimate ground of our existence, available to all, capable of ennobling by its majesty the life of every man who opens himself. It is participation of this order—experiental and not merely political—that alone can guarantee the dignity and autonomy of the individual citizen.[25]

FOOTNOTES

1. Stephen Grabow and Allan Heskin, 1973, "Foundations for a Radical Concept of Planning", *Journal of the American Institute of Planners*. 39(1), p. 108.
2. This so-called ecological literature is immense and includes the works of : Paul Ehrlich, Anne Ehrlich and John P. Holdren, *Human Ecology* (San Francisco, Calif : W.H. Freeman and Company,

1973 : Barry Commoner, *The Closing Circle* (New York, N.Y. : Bantam Books 1974) ; and Gregory Bateson, *Steps to an Ecology of Mind* (New York, N.Y. : Ballantine Books, 1972) ; as well as countless others dealing with environmental and ecological issues.

3. John Friedmann and Mike Douglass, "Agropolitan Development : Towards a New Strategy for Regional Planning in Asia", (Nagoya, Japan : U.N. Centre for Regional Development, 1975), pp. 1-2. Also see Harry W. Richardson, *Regional Growth Theory* (London, England : The Mcmillan Company, 1973), especially Chapter 9, for a perceptive discussion of the relevance of existing and emerging theory for policy purposes.
4. For a particularly well defined and thought out application of the agropolitan idea see Fu-chen Lo and Kamal Salih, "Growth Poles and Regional Policy in Open Dualistic Economies : Western Thought and Asian Reality" (Nagoya, Japan : U.N. Centre for Regional Development, 1975).
5. Kenneth Ruddle, "The Agropolitan Approach to Regional Planning : A View from the Hinterland" (Nagoya, Japan : U.N. Centre for Regional Development, 1975), p. 50.
6. It should be noted here that there is a complementarity between spatial and non-spatial p'anning. The present paper focuses on aspatial issues and looks to implementation tactics to refine spatial policies.
7. For more detailed criticisms of notions of efficiency and large-scale enterprise see : Michael A. Goldberg, 1975, "On the Inefficiency of Being Efficient", *Environment and Planning A*, 7: 921-939 ; and E.F. Schumacher, *Small is Beautiful* (New York, N.Y. : Harper Torchbooks, 1973).
8. The value of traditional approaches to agriculture and other environmental relations are documented in A.P. Vayda (editor), *Environment and Cultural Behavior* (New York, N.Y. : Natural History Press, 1969). The importance of ritual and ritual regulation of human and domesticated animal populations are also noted in Vayda and in Roy A. Rappaport, 1970. "Sanctity and Adaptation". *IO*, 7 : 46-71.
9. One important measure of self-sufficiency is the energy efficiency of production. Traditional cultures seem to have a clear edge here, particularly with respect to agricultural practices, being totally reliant on sunlight as their source of energy and not being able to tap the reserves of stored sunlight represented by fossil fuels. See Roy A. Rappaport, 1971, "The Flow Energy in an Agricultural Society", *Scientific American*, 225 : 116-132.
10. This point is made convincingly by Herbert A. Simon, 1962, "The Architecture of Complexity", *Proceedirgs of the American Philosophical Society*, 106 : 467-482, where he presents the parable of the watchmakers ; one who assembles watches out of preassembled subcomponents and one who assembles each watch without first assem-

bling sub-components. The sub-component approach is a much more effective strategy.

11. See Goldberg, *op. cit.*, as well as C.S: Holling and M.A. Goldberg, 1971, "Ecology and Planning", *Journal of the American Institute of Planners*, 37 : 221-230, for an introduction to the diversity/stability question. See any basic text on biological ecology for a more detailed discussion of the issues involved. The point being made here is that there does seem to be a relationship between the diversity of the elements in a system and the stability of the system (e.g. its ability to respond to a diversity of disturbances of differing magnitudes).
12. The potential dangers which can arise from successful solutions to problems are discussed at some length in Goldberg, *op. cit.*
13. See Vayda, *op. cit.*, for examples of rituals that preserve the diversity of decision rules. In particular, see Chapter 6, "Divination—A New Perspective", by Omar Khayyam Moore, pp. 121-129.
14. Ruddle, *op. cit.*, pp. 35-50, stresses these points in an example he gives about an experimental program in Venezuela.
15. The term in the literature for this schizophrenia is *dualistic dependency* and is the focus for much of the discussion on what's been going wrong with development planning. "Unbalanced growth" is a parallel notion in the economic development and growth literature.
16. See Goldberg, *op. cit.*, for an elaboration of the virtues of small and slow approaches. Also see D. Braybrooke, C.E. Lindblom, *A Strategy of Decision* (Glencoe, Illinois : The Free Press, 1963).
17. This idea that those affected by change must be innovators of change is a basic element in Mao Tsetung, *Selected Readings from the Works of Mao Tsetung* (Peking, China : The Foreign Languages Press, 1972), especially "Serve the People", pp. 310-312 ; and Han Suyin, *China in the Year 2001* (Harmondsworth, England : Penguin Books, 1967), especially Chaprer 5, "Ideology in Command".
18. Friedmann and Douglass, *op. cit.*, note the importance of rural strategies for development when they observe that by the year 2000 more than half the world's population will still live in rural areas, even if the current rates of urbanization continue. Thus future planning for human settlements must consider both urban and rural settlement patterns.
19. Schumacher, *op. cit.*, documents this many times. In spite of the general acceptance of his and related works on appropriate technology, the recently concluded U.N. Conference on Human Settlements, Habitat '76, still called for massive foreign aid programs as part of the "new economic order". Such perpetuation of reliance on external sources of funds, I am afraid, will continue to hamper local development and long run independence.
20. A very cogent argument has been built around such a notion of "social learning" by Edgar S. Dunn in nis challenging study, *Economic and Social Development* (Baltimore, Maryland : The Johns Hopkins University Press, 1971).

21. Lao Tzu, *Tao Te Ching* (Harmondsworth, England : Penguin Books, 1962) p. 125.
22. Friedmann and Douglass, *op. cit.*, and Ruddle, *op. cit.*, as noted go into some detail on shortcomings present approaches to development planning. Support for their position can be found in Lo and Salih, *op. cit.*, and Anthony J. Catanese, "Frustrations of National Planning : Reality and Theory in Colombia", *Journal of the American Institute of Planners*, 39 : 93-105.
23. Simon, *op. cit.*, on the parable of the watchmaker in particular.
24. Saul K. Padover (editor), *Thomas Jefferson on Democracy* (New York, N.Y. : Mentor Books, 1939), p. 24.
25. Theodore Roszak, The Making of a Counter Culture (Garden City N.Y. : Anchor Books, 1969), p. 265.

THREE

LEO HAWKINS

TOWN PLANNER'S FUTURE ROLES

IN THE modern era planners have gone through three sequential, quite distinct role phases. Firstly they were visionaries. Then they were municipal missionaries (like Baron Haussman) or muscular utopians (like Ebenezer Howard). By the 1930's theirs had become the conventional wisdom, and they waited only for a suitable opportunity to implement it on a wide scale. This presented itself with the ending of the Second World War. Sustained prosperity on an unprecedented scale provided the resources, reconstruction the need, and revulsion from the failures of capitalism the ideology.

In this phase, planners found themselves, or perhaps had built for them, the status of what Foley (1960, p. 212) describes as "the bureaucratic keeper of amenity."

But in the 1960's the climate began to change, and with the award of the Nobel Prize to Edward Hayek in 1974, it had swung quite the other way. Planning was perceived at national policy levels as an evil, not an aid, and planners as amongst the most obstructive bureaucrats. The result was an upsurge of articles probing the reasons for this fall from grace, and suggesting alternative careers for displaced or disillusioned planners (Brine, 1977 ; Milne, 1978, for example).

Australia experienced a similar reversal and became what a leading planner called "a non-interventionist society" (Day,

*Paper presented at the "European Meeting on Applied Urban Research," Essen, Oct. 2-4, 1981.

1980). So, for example, at a special national seminar called by town planning students papers were read warning the convenors of the difficult times ahead. Indeed the theme paper was entitled "Planning Education : a poor return from investment in tertiary education" (Apps, 1977).

For a profession whose business is giving the future form, there was a surprising lack of attempt to form the future of town planning. There was essentially an introverted recoilling from a now hostile environment. The research reported here was an attempt to identify some alternative future roles, and the educational consequences that would need to be implemented if those which planners chose were to be effectively performed. This account commences with brief notes on definition and role theory. For a full account see L. Hawkins "The Supply and Demand for Urban Planners in Victoria in the 1980's (Hawkins, 1981).

Methodology

The data on which the conclusions are based was collected in a series of studies in the State of Victoria, Australia in 1979 and 1980. Postal and telephone surveys, and in depth personal interviews were used.

Definition

The study was in great danger of being sucked down a tautological vortex. To identify future roles requires some understanding of present roles (one approach to the study of the future is to regard it as applied history). To do this requires some understanding of who planners are and what they do. But the fact that there is little agreement on these few basic points was part of the reason for mounting the study. This problem was reduced somewhat by the adoption of the definition of planner used by the Canadian Occupational Classification Manual. This publication's approach is in fact valuable because it does not seek to define occupations in isolation, but in relation to each other.

The manual defines an urban planner as someone who :

— Develops programs, plans for controlled utilization of land and physical facilities in rural and urban areas.
— Compiles, analyses and reports data on : economics, legal, political, sociological and physical factors affecting land use.
— Confers with local authorities, civic leaders, social scientists, lawyers and planning and development specialists to devise and recommend zoning arrangements for land.
— Prepares and presents narrative and graphic program plans to civic and industry authorities.
— Recommends measures affecting land use, public utilities, community facilities, housing and transportation, to control and guide community development and renewal.
— Supervises and co-ordinates the work of technologists and technicians.

(Canadian Dictionary of Occupations, 1971, p. 98.)

Role theory

Role has two elements : norms and expectations. Of these, the former are externally determined by, for example, the job situation, and the latter are largely determined within the individual.

Norms are the result of interactions in the environment which produce job requirements and specify levels of performance. Expectations are paradoxically largely the result of interaction between experience on one hand and values and beliefs on the other. The actual role an individual or any organisation plays at any one time or has vailable to it is the result of the interaction of the foregoing elements. They have been pictured in one model by Gowler and Legge (1972) as shown in Fgure 3.1.

The distinction between internal and external influences is important in role analysis, especially when considering future roles, for it enables the identification of the source, and hence some idea of the nature of the dynamics at work. If the environment is weak and the occupation strong there will probably be little change ; the same is true where both are

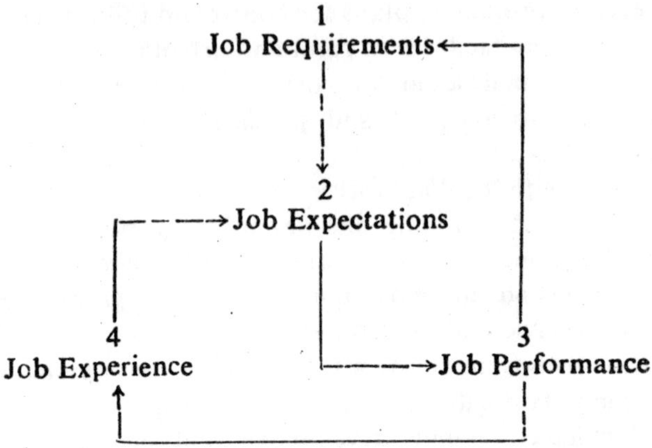

FIG. 3.1 : *The Gowler and Legge Role Model*
Source : Gowler and Legge (1972, p. 15).

weak. If the environment is strong and the organisation weak, there may well be considerable change, but not of the type sought by the profession. The most productive situation is where both the environment and the occupation are strong.

The important method in this study was to identify future role requirements and performance levels as seen by users of planning and compare these with the expectations and experience of individual planners.

The future roles

The 1979 mailed questionnaire attempted to look further than the numbers and to assess the qualitative nature of future demand for planners. It found that a shift towards local government would mean a demand for more technically trained graduates, and fewer "social planners". Paradoxically, there appears to be a continuation of the widespread "upwards drift" of courses for planners, away from threshold skills, towards conceptual, theoretical knowledge. The main source of information on role was, however, the 1980 series of in-depth interviews. Two commonly held views emerged, and may be summarized here, First, that planners should maintain their role as advisers on the allocation of land, but to meet future requirements they must also improve their competence with economic issues and resource constraints, and they must become better communicators, facilitating exchange between competing interests. Secondly, they must become integrators within the urban management context, and this will require them to develop new human relations skills. This emphasis on the importance of communications skills is also found widely overseas. Foley (1979, p. 249) writes :

> "There needs to be a personality change in the planner—a change from the skilled expert to the participatory, collaborative, listening and questioning coordinator. A person who can work equally with politicians and professionals . . . The planner now needs to be a manager over a wide spectrum of public sector effort."

The results were subsequently found to accord with the view of two external observers of urban planning. Writing fourteen years earlier Meier and Duke said :

> "There is a time in the evolution of every discipline when its milieu and the phenomena with which its practitioners must deal, grow in complexity, and responsibilities multiply beyond the capacity to cope with them. That time is at hand for planners" (Meier and Duke, 1966).

They went on to point out that increasing complexity presses a profession toward simultaneous development in two quite different directions : one, specialization of technique and subject matter into slightly overlapping cells of sufficient simplicity for individual decision-making ; and the other, the elaboration of new models and devices for achieving an overview. In Victoria it emerged that this role polarisation was already evident by 1980. The users of planning (employers and colleagues in multidisciplinary teams) emphasised their role requirements as being in specific skills such as report writing, and hard technology such as precise legal advice and economic evaluation. Their performance expectations in these areas were relatively high.

By contrast, the planners themselves had expectations of achieving commanding heights of the overview of policy analysis and qualitative impacts on the lives of urban dwellers. One observer of the planning scene in Australia has detected that planning's new rationale is social policy ; today it has as its central ideology "to make people happier" (Mullins, 1976, page 260). It is planner's experience which has made them more at home in what might be described as the "social sciences" with all the overtones of "soft" technology which go with that phrase.

The dissonance between these two role perceptions will have to be resolved if planners are again to be valued members of urban management teams.

How can this be done ?

Role based education

In his review of the changes in the nature of town planning education in Britain in the decade 1955-1975, A.H. Thomas has drawn attention to the massive increase in the proportion of academic staff drawn from the social sciences, at the expense of land use and design technology. He concludes his evaluation of his data with these words :

> This (study) provides some confirmation that planning education and planning practice, in Britain, are gradually developing into largely independent careers. Of course the interests of educators and practitioners should always overlap : their challenge for the future will be to work out a satisfactory relationship (Thomas, 1980, p. 76).

Without clear role analysis geared to future requirements, planning educators will be left in an objective-less vacuum. In their situation many may relapse into doing their own thing, or nothing. To avoid this role research is essential.

A further avenue for study would be to examine how programs of formal education, or continuing education, could best be developed for the planner's emerging roles.

This study leaves open the question of whether it is a responsibility of existing practitioners to influence the content of courses in the light of their experience and demands as employers. It is, however, worth noting that Faludi (1978, p. 2), while recognising the validity of educational courses providing for "theoretical reflection", has called recently for them to cater for" . . . a pattern of work reminiscent of today's practice with its predominance of typewriters, photo-copying machines and working-party meetings". Before a profession can go beyond "today's practice" and influence the educational content of courses on which its "future" depends it should undertake studies to predict the future roles required of it, and practitioners' expectations.

These studies should be no less extensive or broad-based or systemic than the one reported here.

REFERENCES

Apps, P. (1977). "Planning education : a poor return from investment in tertiary education", *Polis*, Vol. 4, No. 1.

Brue, J. (1977). "A note on job opportunities in planning : educational programmes and the RAPI". *Polis*, Vol. 4, No. 1.

Canadian Dictionary of Occupations (1971), Vol. 1. Ottawa : Information Canada.

Day, P.D. (1980). "Public attitudes to planning in a non-interventionist society", *Planning and Administration*, Vol. 7, No. 1.

Faludi, A. (1978). *Essays in Planning Theory and Practice*. Pergamon.

Foley, D.C. (1960). "British town planning : one ideology or three ?" *British Journal of Sociology*, Vol. 11.

Foley, J. (1979). "Opinion", *Royal Australian Planning Institute Journal*, Vol. 17, No. 4.

Gowler, D. and Legge, K. (1971). "Occupational role development : Part 1", *Personnel Review*, Vol. 1, No. 2.

Howkins, L.F. (1981). *The Supply and Demand for Urban Planners in Victoria in the 1980's (with special reference to local government)*. Royal Melbourne Institute of Technology, Local Government Papers Series.

Meier, R.L. and Duke, R.D. (1966). "Gaming Simulation for Urban Planning", *Journal of the American Institute of Planners*, Vol. 32, No. 1.

Milne, R. (1978). "Alternative jobs for planners ?" *The Planner*, Vol. 64, No. 3.

Mullins, P. (1976). "The physical environment professions", *The Professions in Australia*, ed. Boreham, P. *et al.*, Queensland University Press.

Thomas, A.H. (1980). "The education of British town planners : 1965-1975". *Planning and Administration*, Vol. 7, No. 2.

FOUR

WILLIAM SHALINSKY

THE RELATIONSHIP BETWEEN SOCIAL AND PHYSICAL PLANNING
A Source of Concern

CONCERN with the relationship between physical planning and social planning has waxed and waned over the years. For a period of time in the 1960's there was some interest in this relationship,[1] but it faded in the 1970's. Now, it may be on the upswing again. This increased interest is vital in my opinion because positive contacts between the two must be maintained and increased so that a genuine partnership can be created which will hopefully lead to better quality planning.

I have tried to make three major points in this paper. (1) While a fair amount of contact between physical planning and social planning occurs, it is not always of a constructive nature. (2) This is largely due to insufficient recognition of the factor of paradigm difference, and, (3) to insufficient attention to possible ways of overcoming these differences.

Until these points are seriously addressed, it is my opinion that efforts to develop a constructive relationship between physical planning and social planning will remain elusive.

I will address each of these points in turn and suggest some possible solutions ; but first, by way of background let me explain briefly what I understand by the terms physical planning and social planning.

Physical planning

Physical planning has its focus on the use of land. In

general, it recommends on the uses of (*a*) private land such as housing and commercial development, (*b*) public land such as community facilities, schools, and parks, and (*c*) circulation, that is transportation such as roads and public transit. More specifically, physical planning activities take place in four areas : (*a*) The development of official or general plans to guide the general physical shape of a community. (b) The development of zoning by-laws to provide detail to the official plan. (*c*) Recommendations on the proper division of land. And, (*d*) Recommendations on the design of new development or redevelopment. Increasingly, these efforts have grown to include the use and protection of natural resources as well.[2]

I believe that most physical planning follows a variation of the rational planning process model, which includes six basic steps.

1. Problem recognition, goal formulation and identification of objectives.
2. Studies and forecasts.
3. Plan formulation.
4. Plan evaluation.
5. Plan implementation.
6. Feedback and review.

Although I believe that most physical planning is a variation of the rational planning process, Friedmann and Weaver have noted,

> " . . . some of the major choices that need to be made in organizing the practice of planning ; for instance, whether planning should be :
> Comprehensive or incremental, centralized or developed to local decision units, technocratic or participatory long-term or short-term deterministic or aleatory. These . . . pose the major organizational and methodological issues which must be faced."[3]

Although there are a number of competing interpretations as to what physical planning is at this time, it appears that the above model is still the dominant one.[4]

Social planning

Social planning is the process of alleviating social problems and/or enhancing the well-being of people.[5] It differs from physical planning in that it focuses on the collective needs and concerns of people in relation to five areas.

1. Psychological, including concerns for such things as psychological well-being, loneliness and alienation, and feelings of self-worth.
2. Social, in the sense of a concern with community, opportunities to meet other people and mutual support.
3. Economic, in such areas as poverty or income disparity. It is concerned with people's ability to meet their needs.
4. Political, in relation to power disparities and citizen involvement.
5. Health and safety, which suggests a concern for physical well-being and protection from health hazards and crime.[6]

This form of planning attempts to do its work through four basic areas of activity :

1. The provision of human services. This is the major area of effort in social planning currently. The goals here are : (*a*) the provision of adequate services to meet needs, (*b*) the identification of gaps in services to be filled, and (*c*) the elimination of duplication of services.[7]
2. Locality development or community organization. The major activities of the social planner as community organizer are to encourage self-help and education within the community, a sense of consensus and cooperation among community members, and the development of an organization within the community to carry out various tasks determined by that community.[8]

3. Social planning as advocacy. The basic idea of advocacy is to act on behalf of someone, to represent their interests.[9]
4. Social planning as societal planning. This might take one of two forms—that of broader social programs such as income security programs or the form of social reconstruction, that is, transforming conditions to provide a different social order.[10]

This is my view of social planning. If one were to read others,[11] a number of other ways of looking at social planning would emerge. This lack of agreement as to what social planning is and its potentially enormous scope certainly doesn't make the job of developing relationships with physical planning any easier.

Points of contact between physical and social planning

There are many points of contact between physical and social planning,[12] but in this section I have selected four of these as, hopefully, typical.

1. Inadequate or narrowly focused physical planning encourages social problems, resulting in the need for social planning.
2. Citizen participation in the planning process.
3. The location of human services facilities.
4. The social uses of physical facilities.

I will discuss each in turn.

1. Inadequate or narrowly focused physical planning encourages social problems, resulting in the need for social planning. This is best illustrated through the infamous urban renewal projects of the 1960's when many social problems were created as a result.[13] These projects uniformly resulted in increased need for and programs in three areas of social

planning—community organizing, advocacy and human services. Programs included relocation help, counselling for grief and loss, and financial assistance. More recently, some new community and suburban planning with its emphasis on physical planning has continued to ignore the social planning aspects. As a result, a number of social problems have also resulted for which no services were available.[14]

2. Citizen participation in the planning process. To my way of thinking there are two extremes here : One is where citizen participation is an integral part of the planning process from the beginning.[15] The other extreme is one in which citizen participation is a ritualistic, letter-of-the-law sort of thing.[16] In general, the active support of community organizers, advocacy planners and/or physical planners will encourage more effective citizen participation.[17] When the participation is more integral from the beginning, there will more likely be a cooperative approach.[18] When it is ritualistic, it is more likely to be an adversarial approach between citizens and social planners on the one hand and physical planners and politicians on the other.

3. The location of human services facilities can also cause problems ; some of these are found in the conflict between zoning and halfway houses or group homes, or in the encroachment of social institutions, such as hospitals and universities, on residential areas. Many times, neighbouring residents are suspicious of halfway houses and will fight a zoning bylaw or amendment permitting them. Similarly, various neighbourhoods will attempt to fight the right of large institutions to expropriate. The situation in this section may not only bring social and physical planners into contact, but may well bring social planners with different allegiances into conflict with each other. For example, social planners working as community organizers in a neighbourhood may come into conflict with social planners working with the institution to expand or develop its services and facilities.

4. The final area of contact I want to touch on is in relation to the social uses of physical facilities. Here I am referring to the design of public buildings so they can be used by the disabled and the needs of the disabled in relation

to public transportation and adequate housing. This is an area that has been much neglected until very recently and now seems to be generating a great deal of heat in the often adversarial battles between social planners, physical planners and groups of the disabled who are often militant in pushing their position.

Paradigm differences

The above four points of contact have tended to suggest a more negative, conflictual type of situation. There are of course other situations where a more cooperative type of relationship exists, and I plan to discuss these in the next section. But first I would like to look at some of the difficulties which seem to emerge from the above situations and which appear to get in the way of closer cooperation between physical planning and social planning, and why this appears to be the case. I believe that the concept of "paradigms" is central to this understanding. Kuhn has described paradigms as having two meanings.

> "On the one hand it stands for the entire constellation of beliefs, values, techniques, and so on, shared by the members of a given community. On the other, it denotes one sort of element in that constellation, the concrete puzzle-solutions which, employed as models or examples, can replace explicit rules as a basis for the solution of the remaining puzzles of normal science."[19]

This latter meaning refers to the process of learning a theory through the study of its applications, including practice problem solving. Kuhn refers to this as part of the process of professional initiation.[20] The concept of paradigm conjures up for me a picture of a community or group of people who have developed a certain history and ways of operating that are acceptable and rules as to what is unacceptable. This poses problems for any inter-disciplinary relationships that one may wish to develop. I would like to discuss four of these differences which appear to me to cause much difficulty.[21]

1. Differing perceptions of goals and means.

2. Different language.
3. Differing geographic horizons.
4. Different perceptions as to who the clients are.

1. Differing perceptions of goals and means. As noted in the earlier sections, physical planning tends to focus on land use and physical structures, while social planning focuses more on social needs, concerns and programs. As one person put it, "... urban planners often content themselves by assuming that if properties are cared for, people's well-being will follow suit."[22] In addition, social planners are often trained in schools of social work which operate on a very different paradigm from that of urban and regional planning schools. And on top of that, the organizations for which each works seem to differ significantly. In Canada at least, most physical planners work for some level of government, while most social planners work for either the voluntary or quasi-public sector, including social planning councils, welfare federations, united appeals, and social or health services committees. The situation is further complicated in that some social planners are trained in schools of urban and regional planning and tend to follow the physical planning pattern.[23] And finally, the perceptions of the roles of each differ. As Qadeer notes, "... the stereotypical ideal about the scope of social planning in urban planning is that of a source of appropriate data required by a master planner-designer ... Social planning aspires to go beyond this state. It seeks a more activist role in the planning process ..."[24]

2. Different language. It appears to me that people trained in social planning and those trained in physical planning use some different terms from each other, but often use the same words in a somewhat different way, based on training, indoctrination and frame of reference. Examples of different terms might be zoning bylaw or master plan, on the side of physical planning, and social services or multi-service agency, on the side of social planning. One memorable example of the differential use of the same term is that of "neighbourhood". What is a neighbourhood ? I personally like Gans' comments in which he criticizes the physical planner's typical definition of neighbourhood as one of three to five

thousand people with the usual amenities. Gans comments that this is too large for any but the most superficial relationships. He adds that most people have significant social contact only in the block and with friends outside the block.[25] In addition, physical planners tend to use physical boundaries to divide a community into neighbourhoods, while social planners and citizens tend to see this process more in social and psychological terms.

3. Different geographic horizons. The discussion of neighbourhoods might serve as an illustration of this category as well. But it is also more complicated than that in my view. Social planners may focus on community organization activities on a neighbourhood scale, and social services on a metro, regional, provincial, or state level, or even planning with sectarian units at any of these levels. Physical planners, on the other hand, seem to focus generally on a municipal or metro scale, and have trouble with a neighbourhood scale.[26]

4. Differing perceptions as to who the clients are. Often social planners are concerned with the poor, or with ethnic, religious or radical groupings, while physical planners seem most frequently to operate on the basis of the common good, or more middle-class concerns. I suggest this one cautiously as I recognize that there is a wide difference of opinion in both social and physical planning on this subject. Nevertheless, I think that these perceptual differences do exit.

The above difficulties were highlighted at a seminar for social planners and physical planners, held a few years ago. An eminent Canadian physical planner, when called upon to summarize a number of papers said.

> "In preparation for these remarks I read all the papers very carefully—and it struck me that I was only getting thoughs on social policies that affect the physical fabric of the community, things that land-use planners can put into official plans and design for, from. . .a physical planner.
>
> That's where the problem lies—we're relly simple little characters, we people who plan road systems, sewers, transportation, etc. We can calculate how many fire halls, police stations, and how many schools are likely to

be needed, or how much recreation space might be required (although that is open to question). But, we have absolutely no idea of the physiological effects of various densities and even less idea of the psychological effects, though we do know there are a lot of problems. Now, I would agree with much of what Peter Smith (a Social Planner) has said but I wonder if he has gone far enough in putting things the way councillors, physical planners, and the people affected, can understand?"[27]

Another participant in the same seminar noted that the committee, which had reviewed the Planning Act of Ontario,

"Concluded that social objectives were not appropriate to municipal planning because the available tools relate to physical development.[28]

At the end of the day, it was clear that social planners were pretty silent about what they have to offer physical planners.[29]

Well, what can we do about all these difficulties ? The next section will provide some clues.

Possible ways of increasing positive contact leading to a healthy partnership

In this section I want to focus on two elements which appear necessary in the development of solutions to the problems discussed in the previous section.

1. The formation of new paradigms.
2. Improved methodology.

The new paradigms will hopefully provide the direction and motivation while the improved methodology will suggest the mechanism to make it work.

New paradigms

Kuhn, Perloff and Sherif[30] have all provided clues as to how one goes about selecting new paradigms.

Kuhn discusses the process of transition from one paradigm to another in times of crisis. He states,

> "The transition from a paradigm in crisis to a new one, from which a new tradition . . . can emerge . . . is a reconstruction of the field from new fundamentals, a reconstruction that changes some of the field's most elementary theoretical generalizations as well as many of its paradigm methods and applications."[31]

He comments further that it is necessary to recognize crisis. "As an appropriate prelude to the emergence of new theories. . . "[32] And finally, "The preceding remarks should suffice to show how crisis simultaneously loosens the stereotypes and provides the incremental data necessary for a fundamental paradigm shift."[33]

Sherif, in an entirely different setting, has looked at how one can overcome intergroup conflict. He and his colleagues created an intense intergroup rivalry in a children's summer camp setting. At that point they attempted to reduce the intergroup conflict by various methods. They found that only one system worked effectively—the use of superordinate goals, that is, goals which require the cooperative effort and resources of both groups to achieve and which are seen as appealing and necessary to those involved. These goals encouraged contacts, working together and a broader view of things.[34] Sherif's findings are very similar to Perloff's recommendation that social and physical planning may be better linked through the use of integrating concepts and goals".[35] He describes three of these integrating goals, which fit into Sherif's criteria very nicely. They are :

1. "A decent home and suitable environment for every family.
2. Jobs for all and a minimum family income.
3. Adequacy and equality of public service and facilities."[36]

The message is to rise above the present and proceed to a new paradigm. But how ? Some clues appear in the work of Friedmann and Maruyama.[37]

Improved methodology

Maruyama describes what to do when there is conflict between people with different paradigms.

1. Don't argue or call each others name.
2. Recognize that each party has probably converted what the other said into his/her own paradigm, and as a result, each doesn't really understand the other.
3. It is therefore important to identify the paradigm upon which each party's thinking is built, and into which the statements of the other are converted. This will hopefully permit each party to realize that their paradigms are different, and in what ways.
4. He describes one method of attempting to carry out point three.
   "This process consists of getting into the head (not shoes) of another person. One erases as much of his/her own paradigm as possible from his/her own head, and gets into and thinks in the paradigm of another person. Instead of disagreeing, one tries to think exactly like the other person."[38]

This is not very different from Friedmann's call for mutual learning in the transactive planning style which is a process "with joint exploration of problems and possible solutions to them."[39] Although Friedmann was referring to the expert and client roles, it seems to me that one set of experts sees the other experts in the client role—as only having "personal knowledge" about the first experts' area. The trick will be to share each other's expertise a la Maruyama, leading hopefully to a new paradigm.[40]

A number of examples exist which illustrate the effort to develop new paradigms and the processes by which these might be accomplished. I would like to very briefly mention five of these.

1. Legislation, which requires social and environmental impact assessment, creates the potential for new paradigms as suggested by Perloff and Sherif.[41]

2. The The decentralization of physical planning into local site offices creates the opportunity for intensified dialogue between physical planners, social planners and neighbourhood residents.[42]
3. The Cleveland Policy Planning Report, with its stress on advocacy of the interests of the poorest citizens of the city in social terms.[43] This new form of official plan has gone far beyond the more typical plan in its attention to social rather than physical concerns ; however, the introduction of social goals and policies in the more usual plans are also a significant change.[44]
4. Joint planning, or at least some participation by social planners in physical planning and by physical planners in social planning.[45] The Vancouver social planning department provides one approach.[46] The department was formed in 1968 as a city department partly at the prompting of architects, planners and engineers. As Egan notes, "They recognized that the knowledge explosion plus the increasing specialization in their respective fields of expertise, made it difficult for them to assess fully the social impact of their plans."[47] One of the department's functions is in relation to social environment planning.

   Egan comments,

   "The expression 'Social environment planning' has been adopted by the social planning department to describe its activities in the fields of physical development including zoning, housing, transportation, and social, recreational, and cultural amenities."[48]

   A second approach might emerge through the use of a new paradigm which is called neither social planning or physical planning. Qadeer has suggested two such ideas in his levels IV and V. Level IV would have the tasks of, "integrated comprehensive planning : Social policies and social service delivery programs (would be) components of an area-wide development plan".[49] And level V would have the tasks of redesigning social institutions and creating innovations in social organizations.[50] Both of these would seem to be beyond most planning efforts at this time.[51]

5. It is desirable to encourage more day-to-day contact on interdisciplinary committees. The processes described by Maruyama, Brill, Kane and Lonsdale *et al.*[52] must be used to help the different kinds of planners talk to each other, learn to understand each other, work together, and respect each other so that a partnership might result. From my own experience, the need for this dialogue is essential. I have been teaching in a School of Urban and Regional Planning for several years. During that time, the physically oriented planners and I have been struggling with each other's planning paradigms and have finally worked out an understanding based on the processes described above.

CONCLUSION

I have tried to look at the present state of the relationship between physical planning and social planning as I see it. I have concluded that it is in need of repair. I have suggested some reasons for the difficulties in this relationship, and some of the steps to reach possible solutions. I think Ozbekhan has captured for me the meaning of true understanding between people with different outlooks, and a goal that I think we seriously need to strive for, when he quoted a Zen proverb as follows,

> For the man who is ignorant, trees are trees, waters are waters, and mountains are mountains. When that man gains understanding, then trees are not trees, waters are not waters, mountains are not mountains. And when, at last, he attains wisdom, once again, trees are trees, waters are waters, and mountains are mountains.[53]

And Ozbekhan adds in brackets, "indeed, but with a difference !" I think we have not attained wisdom or understanding yet, but it is to be hoped that we can and will.

FOOTNOTES

An earlier version of this paper was presented at the American Planning Association Conference, October 16, 1979.

1. See, for example, Robert Perlman, "Social Welfare Planning and Physical Planning", *Journal of American Institute of Planners* 32 (July, 1966) : 237-241, Harvey Perloff, "Common Goals and the Linking of Physical and Social Plannings : in *Urban Planning and Social Policy*, eds. Bernard J. Frieden, and Robert Morris (New York : Basic Books, 1968), pp. 346-359, and Canadian Welfare Council, *Integration of Physical and Social Planning* : Report of a Seminar held on April 13-14, 1967 (Ottawa : Canadian Welfare Council, 1967).
2. See F. Stuart Chapin, Jr., *Urban Land Use Planning*, (2nd ed, Urbana : University of Illinois Press, 1965), Arthur B. Gallion and Simon Eisner, *The Urban Pattern : City Planning and Design* (3rd ed. ; New York : D. Van Nostrand, 1975) and Peter Hall, *Urban and Regional Planning* (Middlesex, England : Penguin Books Ltd., 1975). Because of this emphasis on the use of land, I have chosen the term, physical planning. Other synonymous terms are Urban or City Planning, Urban and Regional Planning, Land-Use Planning, and Planning.
3. John Friedmann and Clyde Weaver, *Territory and Function : The Evolution of Regional Planning Doctrine* (Los Angeles : UCLA, 1977), p. 4a.
4. For one overview of the present range and direction of contemporary planning thought, see Thomas D. Galloway and Raid G. Mahayui, "Planning Theory in Retrospect : The Process of Paradigm Change, "*Journal of American Institute of Planners* 43 (January, 1977) : 62-71.
5. This is a broader view of social planning than is usual in the social work, literature, largely because physical/urban planners tend to see the various fields of community organization, community development, social planning and social action as fitting into concept of social planning.

   See, for example, Mohammad A. Qadeer, "The Scope of Social Planning in Urban Planning", *Plan Canada* 17 (June, 1977) : 86-95.
6. These five areas are based, in part, on Abraham Maslow, *Motivation and Personality* (New York : Harper and Row, 1970), Qadeer, 86-95 and League of California Cities, "Social Needs Assessment : A Scientific or Political Process ?" in *Strategies of Community Organization*, eds. Fred M. Cox *et al.* (3rd ed. ; Itasca : F.E. Peacock, Publishers, 1979), 327-337.
7. Andrew Armitage, *Social Welfare in Canada : Ideals and Realities* (Toronto : McClelland and Stewart, 1975) and Jack Rothman, "Three Models of Community Organization Practice, Their Mixing and Phasing" in *Strategies of Community Organization*, eds. Fred M. Cox *et al.* (3rd ed. ; Itasca, Ill. : F.E. Peacock, Publishers, 1979), pp.25-45.

8. Rothman, pp 25-45 and S.K. Khinduka, "Community Development Potentials and Limitations" in Cox, pp. 356-364.
9. Norman Krumholz *et al.*, "The Cleveland Policy Planning Report" *Journal of American Institute of Planners* 41 (September 1975) : 298-307 and 173-318 and Lisa Peattie, "Reflections of Advocacy Planning, *Journal of American Institute of Planners* 34 (March 1968).
10. See Armitage, pp. 16-37 and 197-206 ; John W. Dyckman, "Social Planning, Social Planners and Planned Societies, *Journal of American Institute of Planners* 32 (March (1966) : 66-76 ; John Friedmann, *Planning and the Good Society* (Paper No. 7, Association of Collegiate Schools of Planning, 1974) ; Armand Lauffer, "Social Planning in the United States : An Overview and Some Predictions", in Cox *et al.*, pp. 297-299 ; and Robert Mayer, *Social Planning and Social Change* (Englewood Cliffs, N.J. : Prentice-Hall, 1972).
11. See, for example, Thomas H. Logan, "The Meaning of 'Social' in Planning Curricula : A Review of Courses Offered in Member Schools of the Association of Collegiate Schools of Planning, "*Bulletin of the Association of Collegiate Schools of Planning* 14 (Autumn 1976) : 1-14 and Rothman, pp. 25-45.
12. Qadeer, 86-95.
13. See, for example, Herbert J. Gans, *The Urban Villagers* (New York : The Free Press, 1962) on the West end of Boston and Graham Fraser, *Fighting Back : Urban Renewal in Trefann Court* (Toronto : Hakkert, 1972) on downtown Toronto.
14. Examples include such things as boredom, delinquency, public housing ghettos, and family problems. See Leon Kumove, "Social Planning for New Communities", *Contact* 8 (August 1976) : 164-165. In addition, the continuing process of gentrification, the movement of the middle class into the central city, has revived many of the problems of urban renewal of the 1960's. See, for example, "Symposium on Neighbourhood Revitalization", *Journal of the American Planning Association* 45 (November 1979) : 460-559.
15. See the processes described in John Friedmann, *Retracking America : A Theory of Transactive Planning* (Garden City, New York : Anchor Press, 1973), pp. 171-193 ; John Hitchcock and Donald Bellamy, "Neighbourhood Planning and Neighbourhood Practice", *Plan Canada* 17 (June 1977) : 96-106 ; Harry Lash, *Planning in a Human Way* (Ottawa : Ministry of State for Urban Affairs, Canada, 1976) ; and the higher levels of Arnstein's ladder, Sherry Arnstein, "A Ladder of Citizen Participation, "*Journal of the American Institute of Planners* 35 (July 1969) : 216-224.
16. This is illustrated in the lower levels of Arnstein's ladder, Arnstein, 216-224 and the inclusion of public participation roughly at step 4 of the Rational Planning Process, James J. Glass, "Citizen Participation in Planning : The Relationship Between Objectives and Tech-

niques", *Journal of the American Planning Association* 45 (April 1979) : 180-189.

17. For examples, see the work of Saul Alinsky, *Reveille for Radicals* (New York : Random House, 1969) and Fraser.
18. See Dan Burns, "Neighborhood Planning : A Planner's Perspective", in *Social Objectives and Urban Planning : Proceedings of a Seminar*, June 8-9, 1978 (Toronto : Social Planning Council of Metropolitan Toronto, 1978), Fraser and Lash.
19. Thomas S. Kuhn, *The Structure of Scientific Revolutions* (2nd ed. ; Chicago : University of Chicago Press, 1970), p. 175.
20. Kuhn, p. 47. For further discussion on the subject of paradigms, see Galloway and Mahayni, Magorah Maruyama, "Paradigms and Communication", *Technological Forecasticg and Social Change*, 6 (1974) : 3-32, and Max Siporin, "Practice Theory and Vested Interests", *Social Service Review* 52 (September 1978) : 418-436.
21. These ideas initially came from Perloff, pp. 346-359.
22. Qadeer, p. 94.
23. See, for example, George C. Hemmens *et al.*, "The Practitioner's View of Social Planning", *Journal of the American Institute of Planners 44* (April 1978): 181-192.
24. Qadeer, p. 92.
25. Herbert J. Gans, "The Effect of a Community on Its Residents : Some Considerations for Sociological Theory and Planning Practice", in *People and Plans : Essays on Urban Problems and Solutions*, ed. Herbert J. Gans (New York : Basic Books 1968), pp. 22-23.
26. See Hitchcock and Bellamy, pp. 96-106 and Perloff, pp. 346-359.
27. Max Bacon, "Responses to Presentations", in *Social Objectives and Urban Planning*, p. 175.
28. Allan O'Brien, "Why Social Objectives ?" in *Social Objectives and Urban Planning*, p. 18.
29. A local planner has suggested that, ". . . in many ways, social planners don't have their own paradigm. Often they work from a corrupted version of the social worker's paradigm because the social planning segment of the profession is so small. It's something like the urban planners using the construction industry paradigm." Personal Communication from Ernie Ginsler, Executive Director, Kitchener-Waterloo Social Planning Council.
30. See Kuhn, Perloff, Muzafer Sherif *et al.*, *Intergroup Conflict and Co-operation : The Robbers Cave Experiment* (Norman, Oklahoma : The University of Oklahoma, 1961) and Muzafer Sherif, *Social Interaction : Process and Products* (Chicago : Aldine Publishing, 1967).
31. Kuhn, pp. 84-85
32. Kuhn, p. 85.
33. Kuhn, p. 89.
34. Sherif *et al.*, 1961, pp 151-196, 208-212 and Sherif, 1967, pp. 451-460.

35. Perloff, p. 347.
36. Perloff, p. 350.
37. See Friedmann, 1973 and Maruyama.
38. Maruyama, p. 26.
39. Friedmann, 1973, p. 245.
40. It would be remiss not to mention as well the valuable work in the area of communication in intergroup and interdisciplinary work and the work in systems approaches to social problems. For the former, see, for example, Joseph Luft, "The Johari Window Model", in *Small Group Communication : A Reader*, eds. Robert S. Cathcart and Larry A. Samovar (3rd ed. : Dubuque Iowa : William C. Brown, 1979), pp. 69-78 and Edgar H. Schein, *Process Consultation : Its Role in Organization Development* (Reading, Mass. : Addison-Wesley, 1969). For the latter, see, for example, Russell Ackoff, *Redesigning the Future* (New York : John Wiley and Sons, 1974) and Eric Trist, "Collaboration in Work Settings : A Personal Perspective", *Journal of Applied Behavioral Science* 13 (July-September 1977) : 268-278. Both of these areas are related to this paper, but are separate papers in themselves.
41. See, for example, K. Christensen, *Social Impacts of Land Development : An Initial Approach for Estimating Impacts on Neighborhood Usages and Perceptions*. (Washington, D.C. : The Urban Institute, 1976).
42. See Burns and Hitchcock and Bellamy.
43. Krumholz.
44. For example, see Pater Smith, "Some Initial Approaches", in *Social Objectives and Urban Planning*, pp. 114-41.
45. See Kumove, 164-169 and H.M. Wirz, *Social Aspects of Planning in New Towns* (London : Saxon House Lexington Books, 1975).
46. Maurice Egan, "Social Planning in Vancouver", *Plan Canada* 17 (June, 1977) : 118-126 and Christiane McNiven, "The Vancouver Social Planning Department", in *Community Work in Canada*, ed. Brian Wharf (Toronto : McClelland and Stewart, 1979), pp. 197-224.
47. Egan, p. 119.
48. Egan, p. 123.
49. Qadeer, p. 94.
50. Qadeer, p. 94.
51. But for examples approaching these levels, see Krumholz and Lou D'Amore and Sheila Rittenberg, "The Saint John Human Development Project—A City Shaping Its Future Through Public Participation", *Urban Forum* 3 (August-September 1977) : 4-15.
52. Maruyama, pp. 3-32, Naomi Brill, *Teamwork : Work Together in the Human Services* (New York : J.B. Lippincott, 1976), Rosalie A. Kane, *Interprofessional Teamwork*, Manpower Monograph Number

8 (Syracuse. N.Y. : Syracuse University School of Social Work, 1975) and Susan Lonsdale et al., *Teamwork in the Personal Social Services and Health Care* (Syracuse, N.Y. : Syracuse University School of Social Work, 1980).

53. Hasan Ozbekhan, "The Emerging Methodology of Planning, "*Fields Within Fields Within Fields* 10 (Winter, 1973-74) : 63.

FIVE

IAN MASSER

THE ANALYSIS OF PLANNING PROCESSES
A Framework for Comparative Research

INTRODUCTION

DURING the last ten years there has been a remarkable growth of interest in the study of planning processes of all kinds from a variety of standpoints. This is evident in the great diversity of empirical studies which range from overall descriptions of planning processes to more detailed examinations of constituent elements such as plan formulation and policy implementation. In the course of these studies a variety of subject matter has been investigated which includes both narrowly-focused local planning projects and broadly-based schemes of national legislation. These studies also reflect very different academic and professional standpoints which range from a primary concern with theoretical models of political power and social structure to more limited technical investigations leading to recommendations for improving the efficiency of organisational performance.

This diversity of interests has brought with it many problems which can be seen in the ambiguity and confusion that exists regarding the terminology that is used in these studies. These problems are accentuated by the general lack of communication between academics and practitioners which reflects

Paper presented at the "European Meeting on Applied Urban Research", Essen, Oct., 2-4, 1981.

fundamental differences of standpoint. In addition, the widespread use of the case study approach in this field has produced its own methodological problems in respect of transferability and comparability, particularly where there are great differences in the substantive nature as well as in the treatment of the subject matter that is being investigated.

These problems are particularly marked in the field of urban and regional planning where there is a relatively long-standing interest in the study of planning processes which is associated with a professional interest in improving performance.

Because of this, there is a pressing need to identify research priorities for the study of planning processes which take account of both the traditional concerns of practitioners and recent developments in academic policy analysis.

A scheme which identifies some key features of planning situations is outlined in this paper and its implications are discussed in relation to the findings from a number of case studies of planning processes. The choice of case studies is governed partly by the author's own experience and partly by the principle of maximum diversity. Consequently, the selected case studies range from analyses of planning processes as a whole to investigations of the implementation of previously agreed policy recommendations. As a result of this analysis, a revised scheme is put forward as a basis for comparative research.

A framework for classification

Most writings on planning theory do not go into great detail with respect to the criteria that might be used with respect to the choice of case studies. A partial exception to this rule can be found in the work by Bolan and Nuttall (1975) who selected four case studies in large metropolitan centres where a full range of complexity and diverse political behaviour could be observed to test the model set out in an earlier paper (Bolan, 1969). Three of these case studies concerned efforts to implement broad-scale, comprehensive, systems-oriented plans, each of which was on the brink of a crucial act of implementation which would irreversibly carry forward the purpose of the plan. These were the Boston inner belt expressway scheme, the Lower

Manhattan expressway project and the Pittsburg great high schools plan. The fourth case study concerned the development of an affiliated hospital centre in Boston. This was regarded as being no less complex than the others in terms of decision processes and the levels of interaction that occurred amongst actors but the scope and scale of its issue attributes and the limited character of the planning methods that were employed made it different from them.

On the other hand, van Meter and van Horn (1975) suggest that policies can be classified according to two distinguishing characteristics : the amount of change that is involved and the extent to which there is goal consensus amongst the participants involved. They argue that effective implementation is most likely to take place when the amount of change involved is marginal to the existing situation and when there is a general consensus among those involved regarding goals. Conversely, effective implementation is likely to be most difficult whenever substantial changes are involved in situations where there is no consensus regarding goals.

Both these sets of criteria are incorporated in the classification of planning situations that has been put forward by Bryson and Delbecq (1979, p. 168) as a basis for their study of contingent planning behaviour. A basic distinction is made in this classification between easy-to-deal with and difficult-to-deal with planning situations in terms of their goal dimensions and five key features of planning situations are identified according to these two categories. These concern the number of groups that is affected, the degree of value agreement amongst participants, the implications for organisational structure, the implications for resource allocation and the technical questions involved respectively.

From this scheme it can be argued that the complexity of planning processes increases as the proportion of difficult-to-deal-with components increases. The simplest type of planning situation, then, is that which occurs when all five features fall in the easy-to-deal-with category. The most complex planning situation occurs when all five features fall into the difficult-to-deal-with category. Given the present state of the art, it seems unlikely that a more elaborate classification scheme is needed for practical purposes. It is more useful, instead, to assess the usefulness of the proposed classification scheme in terms of the findings of a limited number of case studies.

The opening and closure of primary schools

The importance of the distinction between easy-to-deal-with and difficult-to-deal-with planning situations is confirmed by the findings of an analysis of the planning environment related to decisions to open and close primary schools in urban areas (Masser, 1980a). The case for a new school is a good example of an easy-to-deal-with planning situation. It involves the following arguments : growth can be expected in the local population ; this will lead to overcrowding in existing schools : when the scale of growth goes beyond a certain point it will be necessary to build a new school. On this basis the education authority can justify the expenditure of its resources and these arguments are likely to be accepted in principle by the two groups involved in the discussions, parents of the (potential) school children and the school teachers, because in both cases there are tangible benefits arising from the planning decision. It is attractive to local parents because their children will not have to go as far to school and they will also be taught in smaller classes. It is also attractive to teachers, to whom it offers the prospect of better teaching conditions in existing schools and possibly new job prospects in the new school. Consequently, once the financial implications of the decisions have been accepted by the education authority, it is in the interests of all parties to push the matter through as quickly as possible and to restrict further discussion to technical details that remain to be resolved.

Planning situations of this kind fall into the easy-to-deal-with category in relation to all the elements that are set out in the scheme. Because they involved a limited number of groups where there is a high level of value agreement, the proposals involve no changes in organisation and can be handled by existing organisational procedures which fall within the existing technological tradition.

Although, on the face of it, the school closure planning situation seems to present similar features, it also contains elements that fall into the difficult-to-deal-with category. The essential argument for the closure of primary schools from the point of view of the education authority is that it will save public money. However, the savings to the education authority must be offset against the additional costs perceived by parents and teachers, and it is in the interests of both parents and teachers to delay the decision as long as possible, because postponement means a continuation of their preferred situation. In situations of this kind, then, there is an obvious conflict of interests and priorities which bring them into the difficult-to-deal-with category in terms of the value agreement component of the scheme. Under these circumstances the formulation of the problem is likely to be challenged in the hope that other solutions will emerge. For example, questions are likely to be asked as to why this particular school was chosen and whether there are not stronger candidates for closure, and so on, which necessitate a constant revision and re-evaluation of the technical evidence before a final deoision can be taken by the education authority (Ferguson and Brown, 1980).

A scheme for town expansion

The school closure case illustrates some facets of difficult-to-deal-with planning situations where there are conflicts of interest between the parties involved. Like Bolan and Nuttall's case studies, it deals with only a single planning episode leading up to a major planning decision. In contrast, Masser's (1980b) study of the planning processes involved in the construction of a town expansion scheme at Smitsveen in Soest in the Netherlands during the late sixties indicates the extent to which this simple classification scheme can be applied to the analysis of multiple decision situations extending over long periods of time. The reasons governing the choice of this case study show a close correspondence to the criteria that are set out in the scheme to describe easy-to-deal-with planning situations :

> "The Smitsveen town expansion scheme must be regarded as a relatively straightforward planning problem in that it was principally concerned with housing construction on a green field site. In projects of this kind there is usually a broad measure of agreement from the outset amongst all the parties concerned as to the nature of the desired outcomes

> . . . the Smitsveen project is also straightforward in the sense that it is typical of hundreds of other projects that were undertaken in the Netherlands during this period. Town expansion schemes of this type constitute the main vehicle for new development in urban areas in the Netherlands and follow clearly-defined formal procedures and well-established codes of practice."

In terms of the scheme then, Smitsveen involves a limited number of actors with clearly-defined roles in a planning situation where there is general agreement regarding the definition of goals. There are also few conflicts with existing value systems in terms of either awareness, priority or intensity of concerns. Its solution requires no changes in existing patterns of resource allocation or organisational structure and utilises techniques which fall well within the existing tradition. Nevertheless, the detailed outcome of the Smitsveen project is not what might have been expected by an outside observer. The study indicates the extent to which the explanation of the apparent paradox that exists between expectations and outcomes lies in the structure of the negotiations that took place between the agencies involved and the way in which a commitment to a particular course of action emerged from these discussions.

A more detailed examination of the decision process surrounding the Smitsveen scheme shows that it consists of a number of distinct episodes, each of which leads up to a key decision point. These can be seen as thresholds which are accompanied by a marked increase in the extent of commitment to a particular course of action. Because of this, the decision process for Smitsveen can be sub-divided into four parts related to the development of outline proposals for town expansion, the preparation of more detailed proposals for the site, the negotiations surrounding the construction of the fully state-subsidised housing, and the negotiations related to the construction of other types of housing respectively. Each of these phases is broadly comparable to the episodes that were studied by Bolan and Nuttall and could in itself form the subject of more detailed investigation.

The findings of the Smitsveen study also suggest new criteria hat might be added to those included in the scheme. They point to the existence of a different kind of threshold which arises as a result of the impact of external factors on the decision process. Two events of this kind occurred in this case. The first concerned the decision by the Province to reject the local authority's preferred scheme on another site which led to a change in the priority given by the local authority to the Smitsveen scheme, and the second concerned the decision of the local authority to take advantage of a new government subsidy which was designed to encourage industrialised construction schemes in order to reduce the financial risk it would incur in the project. Both these events altered existing perceptions of the planning situation and gave rise to changes in the pace and direction of the implementation process. In the case of Smitsveen, their main consequence was to reinforce existing tendencies and to speed up the implementation process. In other circumstances, however, they can have the opposite effect and provide a plausible reason for a retreat from a high to a lower level of commitment. This occurred, for example, in another study of town expansion where a gradual build up of resistance on the part of local residents to the proposed built form culminated in the abandonment of the approved scheme and its eventual replacement by an entirely new set of proposals (Masser, van Hal, Post and van Schijndel, 1978, pp. 143-146).

There is also evidence from other studies in the field of decision making (see, for example, Mintzberg, Raisinghani and Theoret, 1976) to indicate the extent to which the sequence of events is influenced by external factors, especially when complex decision processes extending over long periods of time are involved. The extent to which external factors fall into the easy or difficult categories depends on the extent to which they reinforce or counteract existing tendencies. In Smitsveen, then, the impact of external factors would come into the easy-to-deal-with category in that they reinforce existing tendencies to specificity, whereas in the other case they would come into the difficult-to-deal-with category in that they lead to a reduction in the degree of specificity.

The designation of new and expanded towns

The cases discussed so far involve particular project situations and it is necessary therefore to turn to Levin's (1976) study of new town and town development designation processes to examine the extent to which the project model is applicable to other kinds of planning situation. This is not concerned with the planning process as a whole, but only with the limited set of decisions and negotiations which led up to the formal designation of schemes for a new town in Central Lancashire and a major town expansion at Swindon. These two case studies are broadly comparable to the first stage of the Smitsveen project but they differ from it in that new town procedures are initiated by central government, while town development schemes necessitate a formal agreement between the authority 'exporting' overspill population and the authority who is to receive this population as a result of town development.

The findings of this study suggest that even this limited planning process can be sub-divided into stages ending in thresholds of the kind observed at Smitsveen. In both cases, Levin notes that there is an inception stage which culminates in the commissioning of special studies. This is followed by a draft planning stage which is completed by the approval and publication of formal proposals for designation and the process as a whole is rounded off by a formalisation stage which ends in the final decision by the Minister to approve (or reject) the designation order.

The new town designation case represents an easy-to-deal-with planning situation although many groups are involved, because the Minister, because of the powers vested in him, is both promoter of the scheme and the ultimate decision maker. Under these circumstances, Levin describes the planning process as a 'bandwagon' situation where decisions based on a quite limited range of information and interest generate a degree of commitment which severely restricts the effect of later inputs. In situations of this kind he notes that there is a tendency for those involved to change their goals in order to justify courses of action to which they are already committed rather than to consider changes in that course of action.

On the other hand, the Swindon town development designation process demonstrates some of the features of more difficult-to-deal-with situations. The Minister's position in this case is only as the ultimate decision maker whose decision is needed for agreements that have alrealy been negotiated by the receiving authority and the exporting authority. This is described as a 'tripod' situation where the growth of commitment is relatively restrained by comparison with the 'bandwagon' situation. Under these circumstances, a much wider range of information and interests is brought into the early stages of negotiations and new insights brought to bear late in the process can potentially have a much greater influence on final decisions than was the case with respect to new towns.

The findings of these studies suggest that the distinction that is made in the scheme between a single group and multiple groups should be amended to take account of the differences that have been observed between new town designation procedures, which are characterised by a dominant actor amongst the groups involved, and town development designation procedures, which involve negotiations between parties that can exert little formal influence upon one another. The dominant actor situation can be regarded as falling into the easy-to-deal-with category, whereas bargaining processes of the kind that occurred at Swindon fall into the more difficult category. This distinction is also borne out by the findings of the Smitsveen study where the Municipality of Soest occupied a pivotal position throughout the negotiations involved in the implementation of the project.

The local implementation of central government policy

All the studies described so far concern planning situations which fall mainly into the easy-to-deal-with category. They deal with cases where there is general agreement on desired outcomes, standard procedures are in existence and there is a general acceptance of the technical criteria. Only in the case of the school closure and the negotiations surrounding the expansion of Swindon have conflicts of interest and the lack of a dominant actor introduced elements which fall into the difficult-to-deal-with category. In contrast to these, Pressman and Wildavsky's (1973) study of the activities of the Economic Development Administration in Oakland, California explores some properties of more difficult-to-deal-with planning situations. This was chosen by the authors because of its apparent ordinariness :

"In our study of implementation, we have deliberately chosen case material in which dramatic elements that are essentially self-explanatory are ruled out. There was no great conflict. Everyone agreed. There was only minimum publicity. The issue was not one of over-riding political importance. Essential funds were on hand at the right time. The evils that afflicted the EDA program in Oakland were of a prosaic and everyday character."

TABLE 5.1 : **A Revised scheme of key dimensions of planning situations**

Dimension	Easy-to-deal-with	Difficult-to-deal-with
Level of specificity	High	Low
Group character	Presence of dominant actor amongst participants	Lack of dominant actor amongst participants
Degree of value agreement	No conflict with existing value system in terms of awareness, priority and intensity of concerns	Major conflict with existing value system in terms of awareness, priority and intensity of concerns
Impact on organisational structure	No re-organisation required	Significant re-organisation required
Technical difficulty	Within existing tehnological tradition in terms of comprehension of causation and sophistication of methodology	Radical change from existing technological tradition in terms of comprehension of causation and sophistication of methodology
Impact of external factors	Reinforce previous commitments	Undermine previous commitments

This study concentrates on the processes of implementation that followed the initial setting of goals by central government, the securing of local support and the commitment of both central and local resources. It describes what happened when a new agency that was set up by Congress decided to move into cities with a view to providing permanent new jobs for minorities through economic development and selected Oakland as the location of an experiment to demonstrate how public works and business loans could provide incentives for employers to hire minorities. Despite these favourable conditions and support from city administrators and employers in Oakland, the project showed a marked lack of success in terms of implementation. Six years after its beginning, the proposed construction schemes were only partly completed and the business loan element of the project had been abandoned completely.

The lack of success of the Oakland project is largely due to the extent to which the implementation of the detailed proposals, as distinct from their formulation, involved the co-operation of a large number of agencies whose agreement was needed to avoid blockages to progress. Although these agencies were generally favourably disposed towards the project involved, there were important differences in emphasis and time-scale between them which only emerged during the process of implementation. This was further complicated by the lack of standard operating procedures which meant that a large number of unexpected decisions had to be resolved in the process. Consequently, what appeared to be a basic set of fourteen key decisions involving both the public works and job creation programme aspects of the project was complicated by an additional seventeen unexpected key decisions. In total, these key decision points involved seventy separate sets of agreements with bodies where there was no effective dominant actor.

This study illustrates some of the characteristics of planning situations where difficult-to-deal-with elements predominate. Even so, it should be noted that not all elements in this case fall into the difficult-to-deal-with category. There was general agreement on goals. There were no fundamental conflicts of values and resources were available for implementation. Nevertheless, under these circumstances, the apparently simple became complex because of the insurmountable nature of the organisational problems associated with the project.

A revised classification scheme

Following the discussion of the findings of these selected case studies, a number of revisions must be made to the scheme that was set out earlier for classification purposes. There are incorporated in the modified scheme that is set out in Table. 5.1. From this it can be seen that two completely new elements have been added to the original list of criteria. These concern the impact of external factors, that was illustrated particularly by the Smitsveen study, and the degree of specificity that is involved in the planning situation respectively. Criteria relating to specificity have been included in the scheme because it is felt that a basic distinction can be made between situations where a high level of specificity is obtainable, either through the realisation of a blue-print plan or through some measure of commitment to a formal, irrevocable act, and situations of the kind described in several case studies fall mainly into the former category. These include the formulation and implementation of strategic planning policy of the kind encountered in the land-use transportation studies that were carried out in America during the sixties (Boyce, Day and Macdonald, 1970).

In addition to these two new elements, the criteria relating to the number of groups have been revised to take account of the dominant actor/lack of dominant actor distinction that was noted in the case of the town designation studies. Furthermore, the resource allocation criteria have been incorporated into the organisational structure criteria. This is because most changes in resource allocation automatically imply changes in organisational structure.

CONCLUSION

A general framework for the study of urban and regional planning processes is set out in this paper which points to the need for a much more systematic attack on these problems than has so far been the case if the best use is to be made of case study based research. To mount such an attack a great deal of consultation and extensive collaboration between research workers would be required to develop and implement linked case studies in a comparative framework. The present paper might be regarded as a first step in this direction in that it discusses some of the practical questions that must be resolved in the process. The next step might be to bring together researchers from a limited number of countries who are interested in developing a consistent framework for their work on urban and regional planning processes. It is felt that such a group could usefully take the present paper as a starting point for their discussions.

REFERENCES

Bolan, R.S. (1969) Community decision behaviour : the culture of planning, *Journal A.I P.* 35, 301-10.

Bolan, R.S. and R.L. Nuttall (1975) *Urban planning and politics*, Lexington, Mass.

Boyce, D.E., N.D. Day and C. McDonald (1970) *Metropolitan plan making : an analysis of experience with the preparation and evaluation of alternative land use and transportation plans*, Regional Science Research Institute, Philadelphia.

Bryson, J.M. and A.L. Delbecq (1979) A contingent approach to strategy and tactics in project planning, *Journal A.P.A.* 45, 167-179.

Ferguson, S. and P.J.B. Brown (1980) Issues related to decision making in the planning of inner city primary school provision in the face of declining enrolment *in* W. Buhr and P. Friedrich (eds) *Planning in stagnating regions*. Nomos, Baden Baden.

Levin, P.H. (1976) *Government and the planning process*, Allen and Unwin, London.

Masser, I. (1980a) The limits of planning, *Town Planning Review* 51, 39-49.

Masser, I. (1980b) Some dilemmas of urban planning, *TRP. 27*, Department of Town and Regional Planning. University of Sheffield.

Masser, I., W. Van Hal, W. Post and R. Van Schijndel (1978) The dynamics of development processes : two case studies, *Town Planning Review 49*, 127-148.

Mintzberg, H., D. Raisinghani and A. Theoret (1976) The structure of 'unstructured' decision processes, *Administrative Science Quarterly* 21, 246-275.

Pressman, J.R. and A.B. Wildavsky (1973) *Implementation : how great expectations in Washington are dashed in Oakland*, University of California.

Van Meter, D.S. and C.E. Van Horn (1975) The policy implementation process : a conceptual framework, *Administration and Society* 6, 445-88.

SIX

DIMITRIOS S. DENDRINOS

THE THEORY OF STRUCTURAL STABILITY
Towards A New Paradigm for Planning

THE THEORY OF STRUCTURAL STABILITY : A BRIEF INTRODUCTION

FORMALLY the theory of structural stability emerged as a branch of mathematics in 1972 with the publication of Rene Thom's book translated into English in 1975 as *Structural Stability and Morphogenesis*. This publication ignited a new outburst of work on a topic which was initially developed in the sixties. Referred to as "Catastrophe Theory" (C.T.), the theory of structural stability is an illustrative example of the pure mathematician's choice of words in defining precise abstract concepts. There are terms in the theory ranging from "umbilic bracelets" to "differentiable germs" to "stratified jets" to mention but a few. The term "catastrophe theory" has been cause for considerable confusion and negative impression. According to some, the single most inhibiting factor in the theory's proper acceptance and development has been its name.

Since the appearance of *Stabilite Structurelle et Morphogenese* the complete enumeration of all the publications on the subject is no longer feasible. A few sources are given to guide the interested reader in an exploratory way into the depths of C.T. : the well-known article by Zeeman in Scientific American (1976) popularized the subject. The computer graphics work by Woodcock and Poston (1974) and two recent books by Zeeman (1977) and Poston and Stewart (1978), provide a gradually unfolding view of the theory along with extensive bibliographies.

A number of papers indicate the potential use of C.T. in the social sciences for the non-topology inclined researcher.[1] A good collection of papers can be found in the September 1978 issue of *Behavioral Science* with many suggestive extensions, as well as some criticisms. Some attempts have been made in the area of planning (Amson 1974, Mees 1975, Amson 1975, Wilson 1976, Dendrinos 1976, Poston and Wilson 1977, Isard and Liossatos 1977, Harris and Wilson 1978, Dendrinos 1978 (a and b), 1979.[2] Clearly, we are currently experiencing a mild literary explosion in this area of research, where contributions to the theory itself are as synchronous and numerous as its applications.

What is the theory of structural stability ?

C.T. is a branch of topology, namely differential topology, which has never been the planner's choice of sport ; pure mathematical analysis is normally discounted by planners and policy makers as too rigorous, artificially complex, and lacking realism. Nonetheless, it is the intent of this section to provide the reader with a general exposure and understanding as to the potential usefulness of this new theory.[3] To researchers, whose most sophisticated technique ever used is multiple regression, certainly it is not trivial to show that "diffeo-morphism," the "separation" and "preparation" theorems, the "K-jet of the G-germ" and other such terms have any significant meaning for doing, say, costal zone management. Nevertheless, the reader is asked to keep an open mind during this (quixotic ?) adventure and await for their plangentness to unfold slowly.

Phenomena of rupture, break, thresholds, cycles, collapses, delays, accidents, change of form, and other discontinuities are abundant in nature. Such events, that we shall refer to as "catastrophic", are not only encountered as natural phenomena but also as social events at the micro and macro scale. Families break up, social upheavals and revolutions occur, bureaucracies change their form. Plans go through successions of phases and forms. Land use densities and other urban form variables exhibit discontinuities over space and time. Critical disinvestment decisions that lead to collapses or severe disruption of whole transportation systems are often made. It has become almost a triviality to conclude, as many C.T. proponents have already done and many C.T. critics have been quick to point out with irony, that C.T. is a method for viewing such phenomena ; that it is a new technique filling up a void in the theory of such systems. However, this is not the whole story ; if the claims were to anchor here, they would find the waters shallow. There are other aspects of C.T., within planning theory, that make it intellectually attractive and these angles we will attempt to take here. But in order to reach the Rocky Mountains one has to travel through Kansas, westbound.

To begin with, C.T. is a dynamic theory.[4] It assumes that the behavior of a system at any point in time is under a governing optimizing criterion resulting in the system either being at rest (equilibrium state) or at disequilibrium. In the first case the system can exhibit either stable behavior or unstable one.[5] The theory refers to such optimizing criterion as a "potential function". C.T. through such potential functions associates certain variables referred to as "behavioral" variables (describing the system's behavior) with certain parameters referred to as "control" variables to produce catastrophic events, i.e., continuous dynamic changes in the control variables resulting to discontinuous changes in the system's behavior.

Even with limited statistical background planners have been familiar with regression models where a "dependent" variable is associated with a number of "independent" variables in a linear or non-linear fashion. C.T. provides a more sophisticated association among dependent and independent variables, by deriving the most complex events that can possibly occur in such relationships. It turns out that such events are "catastrophic" and occur at points known as points of "singularities" in the behavior of a system. These points of singularities contain some form of discontinuity in the behavior

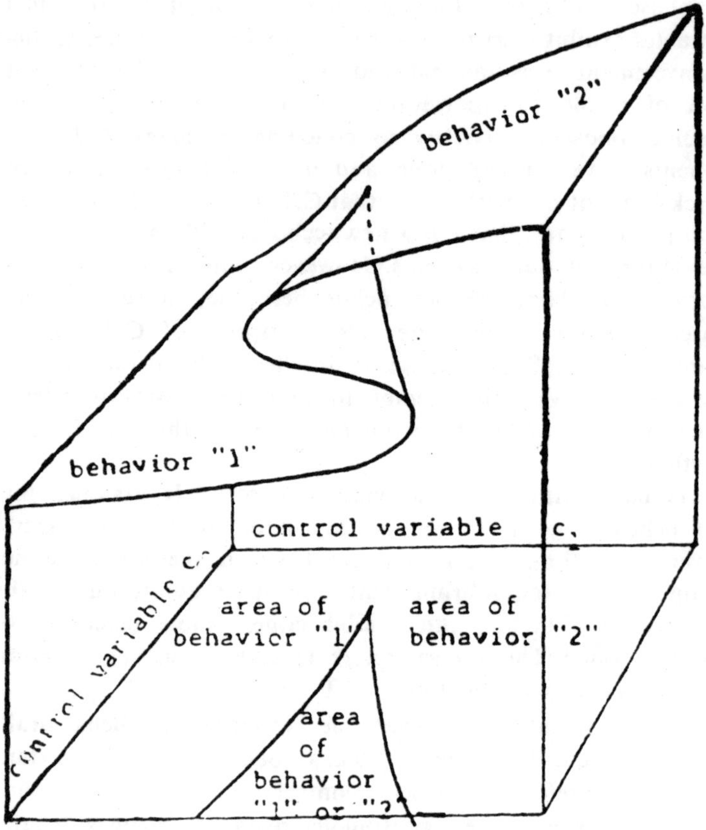

FIG. 6.1 : *The bifurcation of behavior*

of the system under study. Linear and ordinarily used non-linear models are not capable of producing such points.

The derivation of such points of singularities is due to the notion of "bifurcation" that C.T. introduces in the study of systems behavior. Bifurcation occurs when, within certain ranges of the control variables (see Figure 6.1), any given specific set of values of the control parameters could result to a multiple choice of values in the behavioral variables. In other words, within such ranges the behavior of the system is not unique. The ranges of the control variables over which multiple behavior can occur is referred to as the "bifurcation set", and the set's boundaries are the points of singularities, where discontinuities and catastrophic events occur. C.T. claims to describe *discontinuous* events that result from *smooth changes* in the control variables, events that are unexpected without foresight and genuine contributions to the learning of the system's behavior.

Multiplicity in behavior may involve many stable or unstable states. A specific case of two stable and one unstable state of behavior is the cusp catastrophe,[6] shown in Figure 6.2. The equilibrium behavior of the system, measured along the behavioral space (shown in the upper part of the figure), is plotted as the control variables vary (designated as C_1 and C_2 and shown in the lower part of the figure). On that behavioral surface the potential function is optimized and the system is at rest (stable on the surface of the two pleats designated as stable modes, and unstable on the one between them. Within the ranges of the control variables indicated by the dotted part of the control space the behavior of the system could attain either of three equilibrium states (two stable and one unstable). Outside the bifurcation set the system attains a unique stable state. The dotted part is the bifurcation set ; the two border lines are the points of singularities, where discontinuities occur (for example at point E on the path running along the lower sheet there is a sudden jump from A to C, Figure 6.3A ; and similarly there is a discontinuity at point F when moving along the path on the upper sheet, Figure 6.3B).

In the particular case of the cusp catastrophe there are some phenomena in addition to these "jumps" that are of interest. There are events of : "delay" in switching of modes inside the bifurcation set ; "divergence" in behavior (Figure 6.4) ; and finally "hysteresis" (shown in Figure 6.5). C.T. suggests that when anyone of these four phenomena is present in the behavior of a system the observer must look for the other three.

The driving element of C.T. is the possibility of multiple equilibria, the bifurcation of behavior, for certain ranges of values of the control factors. By virtue of the existence of multiple equilibria the theory can infer the properties of the dynamic behavior of the system. The laws of change, both

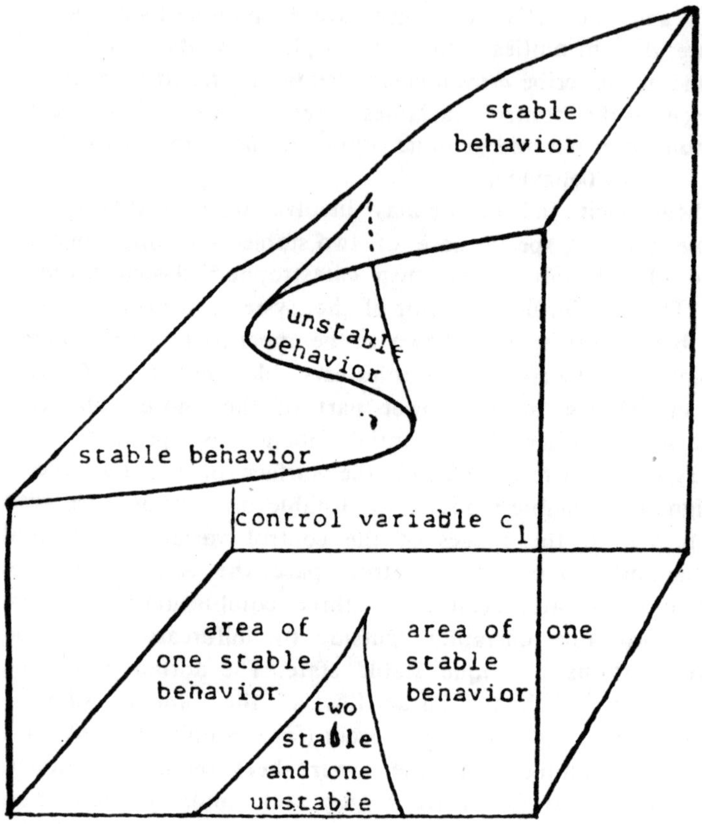

FIG. 6.2 : *Stable and unstable states*

in nature and in society, can be determined by our knowledge of the number of possible (stable) equilibria. Economic analysis is abundant with cases of multiple equilibria. The attitude among scholars in that field, however, has been to dismiss problem formulations that derive multiple solutions as useless and non-interesting for research. It is argued by C.T. theorists that the existence of multiple equilibria is actually of interest in dynamic analysis since it determines genuine dynamic behavior.

In the literature appeared so far on C.T. and its applications there seems to be emerging an important difference

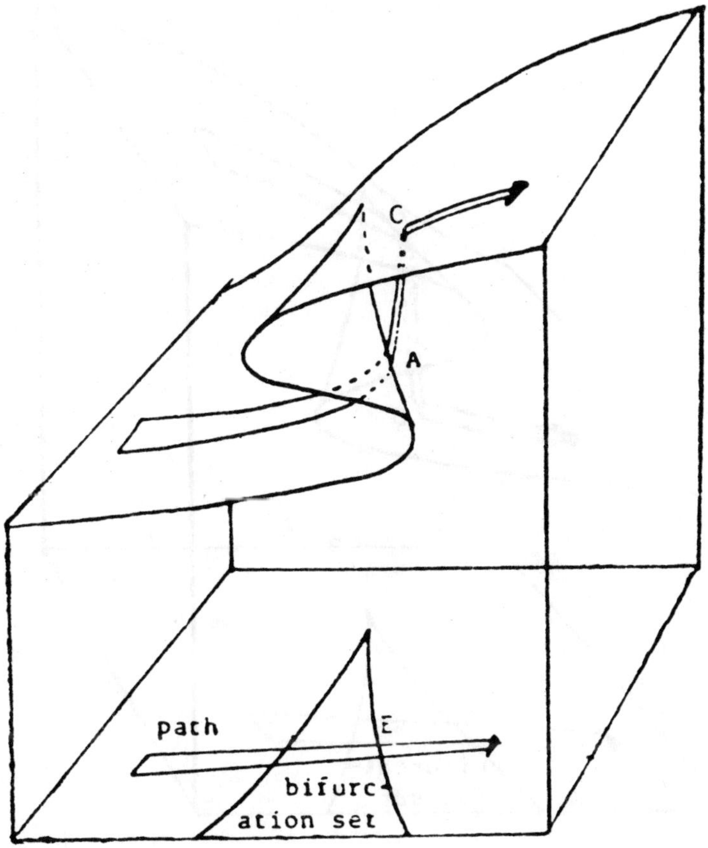

FIG. 6.3 A : *Discontinuity at point of singularity E for a path moving east*

between the so-called "Thom School" of catastrophe theory model building and the "Zeeman School". Zeeman has created models and machines so that he can deduce the catastrophic behavior of a system by postulating its regulating function (the optimizing potential that governs the system's behavior), or its gradient, the laws of change of the system's behavior. Thom, on the other hand, infers the existence of a system's potential by observing directly its catastrophic behavior at a point (locally) and then proposes a model

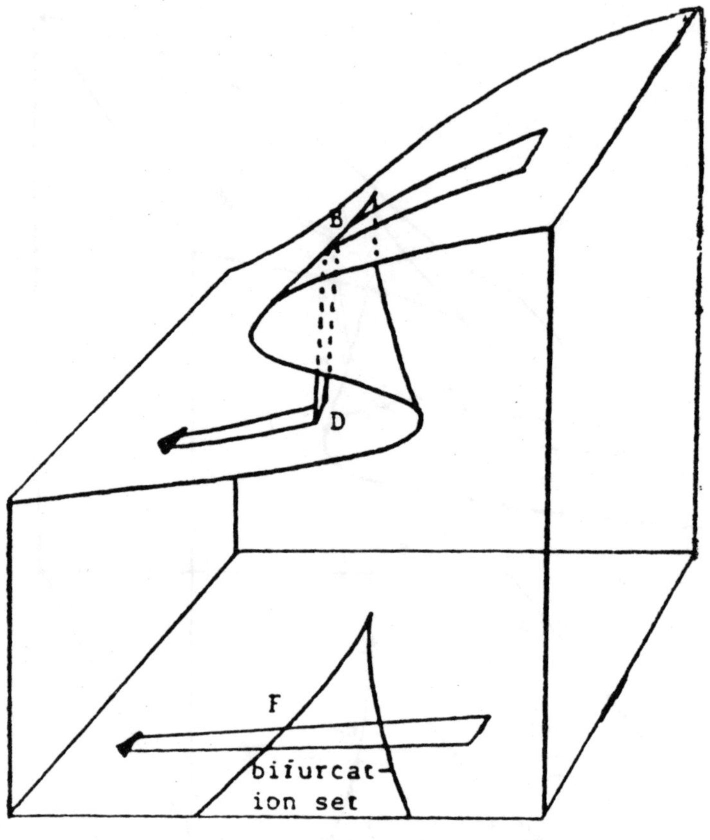

FIG. 6.3 B : *Discontinuity at point of singularity F for a path moving west*

adequately specified (parameterized) so that it would reproduce the locally observed catastrophe so as if it were to govern the system's behavior there.[7] Thus, Thom infers behavior by observing the morphology of a system in a space-time framework.

Both approaches have significant repercussions on our modes of viewing systems and profound effects on our roles as planners and policy makers. Zeeman's model building allows a *global* examination of the system. Thom's model building concerns mainly the *local* properties of systems and

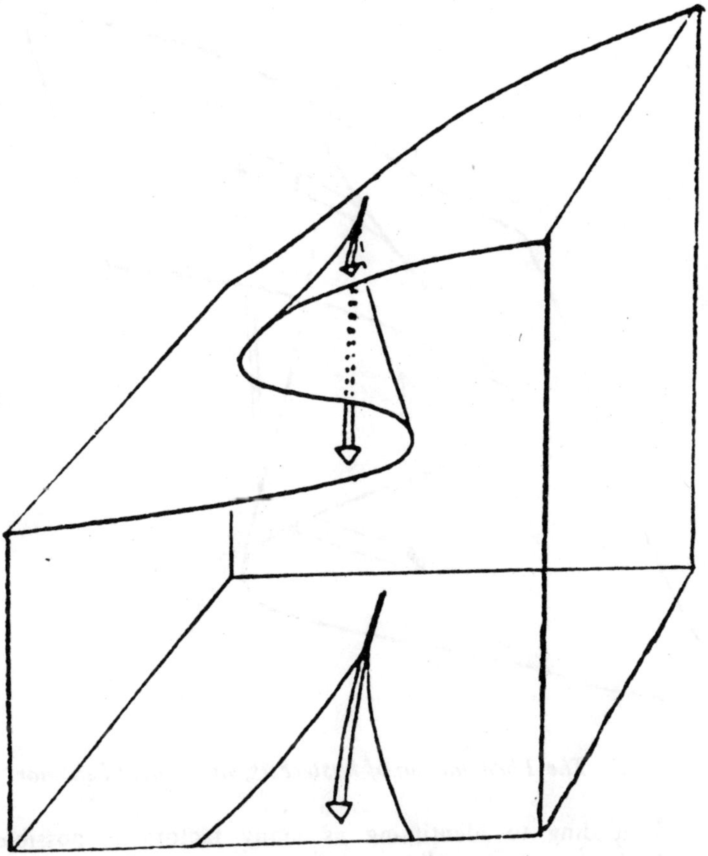

FIG. 6.4 : ***Divergence of behavior for a path inside the bifurcation set***

then it attempts to infer statements on its global behavior from the local observations. Caution should be exercised here. Since not all of Zeeman's work falls under the category bearing his name and since some of Thom's reasoning does not exclude the modeling a la Zeeman, assigning these labels to either must be taken loosely. This does not, however, invalidate the distinction made between the approaches of the two schools.

Researchers interested in developing models to describe, explain and if possible predict events typically use a mode of

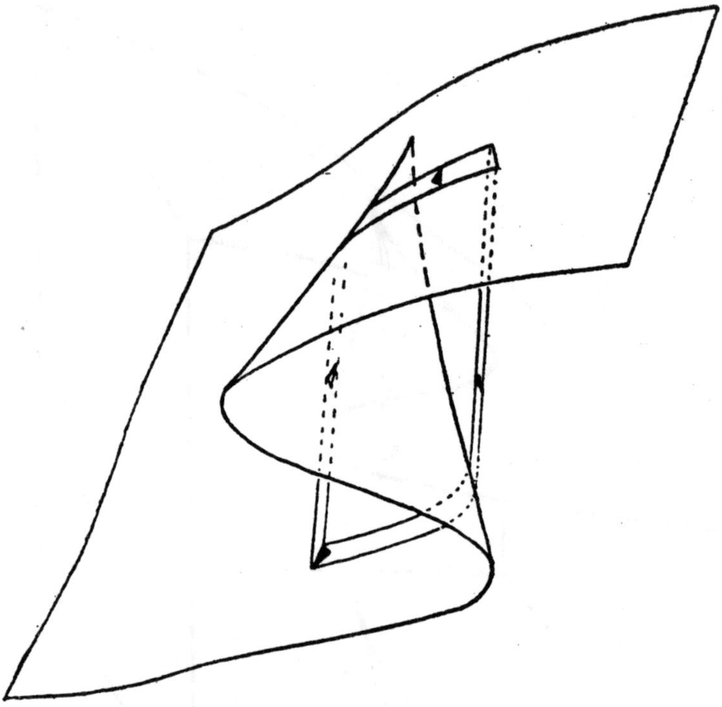

FIG. 6.5 : *The Phenomenon of hysteresis, or cyclical behavior*

analysis leading to identifying as many factors as possible and their interrelationships, which are perceived as bearing on the final outcome. According to Thom's classification theorem a radically different alternative emerges. A system exhibiting morphologically one or any combination of the phenomena mentioned earlier only needs a few key variables for modeling among the numerous background ones. Further, the type of catastrophic event exhibited by the system ascribes a specific number of possible stable states (modes) which the system can attain, which in turn determines the number of the key control variables needed to describe the system's behavior.[8]

Such a theorem is far reaching. It implies that, if proven empirically relevant after scrutiny and rigorous testing, our current way of understanding and theorizing may not be the most effective one. In the social sciences the attitude has been to account for a large number of factors which are perceived to affect the behavior of the system under investigation.[9] C.T. provides a framework for abstracting a system's behavior by looking at the system's configuration (i.e., its macro form) rather than zeroing in the system's micro structure.

The distinction between the Zeeman and Thom schools of model building allows for classifying C.T. models as behavioral or structuralist in nature. Structural models are more descriptive than explanatory, whereas behavioral ones contain more predictive power. According to the Zeeman school, a behavioral model involves stating a problem's potential function[10] (or gradient) defined over a specific set of measurable control parameters affecting a concrete set of behavioral variables so that the problem statement can reproduce the catastrophic event. Such problem formulation falls under the constrained (or unconstrained) non-linear optimization theory. On the other hand, in Thom's structuralist model, the emphasis is different. The starting point is the catastrophic event itself. Given the form of the system's catastrophic behavior, a class of functions is identified which belongs to a given pre-established form (potential function) which in turn corresponds to a given pre-existing set of elementary structures governed by this family of similar, neighboring functions. Thus, starting from the catastrophic event itself, the governing potential of the system is determined at the neighborhood of the behavior where the event occurred. This dichotomy in model building, the deductive method of Zeeman and inductive method of Thom, belong to two alternative scientific paradigms (Dendrinos, 1979). Both schools use identical theorems but the philosophical bases and implication are different. Is there a catastrophic phenomenon present in the structure of scientific paradigms ?

The exposition of C.T. will end here, in order to enter the main scope of this paper ; it is expected that by now the reader's interest has risen enough to seek the primary references and obtain a more comprehensive understanding of the more fine notions of C.T. that are necessary for a full appreciation of its potential use. As an author noted : "without the details there can be no delight."[11]

CATASTROPHE THEORY : A TECHNIQUE IN PLANNING

What are some of the planning issues which C.T. could begin to address ? In the land use sector, for example, if a zoned area has been continuously undeveloped and then suddenly develops with no change in zoning ordinances, then such a discontinuous growth pattern is subject to C.T. formulation and examination. Such developmental events convey much information about land use behavior within metropolitan areas thereby making land speculation a topic amenable to C.T. modeling. For another example, adjacent locations within metropolitan areas have been shown to follow quite different development patterns in certain instances. In the U.S., cases are recorded where adjacent downtown residential areas with identical socio-economic and housing stock characteristics at one point in time, gradually diverged in their dynamic paths, one becoming a deteriorating neighborhood whereas the other continued to grow under identical control environments (location and land use values). Boom towns arising overnight and ghost towns abandoned at equal speed are well documented events. All of these are characteristic of behavior amenable to C.T. analysis.[12]

Numerous authors have been engaged in applying C.T. to planning but their attempts have been isolated and rather impromptu. No concerted effort has been made to develop a problem design with the theory enabling us to create models—an example being the theory of mathematical programming. Obviously, this is neither an appropriate forum to attempt this methodically nor a suitable place to propose an agenda for research by identifying planning topics suitable for C.T. analysis. It is only natural that topics which exhibit catastrophic behavior more obviously than others will be studied first. Wilson and the author simultaneously and separately saw a modal choice model when they first came across the palpable cusp ; there was an easy analogy to be made. The modal choice model in the travel demand part of the urban transportation planning process is an illuminating example of C.T. as applied to transportation. The case is presented from a variety of points of view in Wilson (1976) and Dendrinos (1976) where the specific advantages of C.T. versus the current random choice based dominant theory used to describe modal choice behavior are discussed ; in summary, the four basic elements of a cusp catastrophe (hysteresis, delay, divergence, and jump) that are observed in everyday tripmaking behavior with regard to modal choice are not reproduced by random choice models. Further, the random choice results are only a part of the general results obtained by using the cusp catastrophe model.[13] Higher order catastrophes, which requires more dimensions to plot the bifurcation set are more difficult to perceive visually, and thus more difficult to transfer to specific problems. Recent work by the author and others has attempted to formulate the problem of evolution of cities within a C.T. framework, Dendrinos and Mullally 1979. Although the work remains of a speculative level, empirical research is currently underway to test it.

Economic analysis is an area where C.T. has the potential to make a significant contribution. Models which have been used by planners and other policy makers are classified as "positive" and "normative". Positive models refer to competitive market equilibrium allocations that are Pareto efficient in the sense that no relocation of resources can be found that would make some better off without harming at least one individual. Normative models are various optimization problems where socially optimal allocations are derived. Positive models with dynamic adjustments explicitly stated in the form of gradient functions (versus cases in which only the long-run static equilibrium is sought), can be viewed under C.T. as special cases of optimization formulations. Since such an optimizing criterion (potential function) was not present from the beginning the contribution is to provide evidence for its existence and force the researcher into formulating and interpreting it. For an example of such a dynamic market behavior model the reader is referred to Dendrinos (1979). C.T. provides an opportunity for a *unifying theory of positive and normative models.*

Existing theories based on calculus and modeling continuous change fail to account for discontinuities in nature and society. Discontinuities should not be construed merely as "aberrations" or special cases of existing theories. Rather, they should be viewed as *anomalies* encountered from the application of existing theories or paradigms into the study of natural or social systems. The existence of multiple stable local optima now can be used in dynamic analysis to examine (catastrophic) phenomena previously unaccounted for.

Finally, one of the most heralded properties of C.T. it its capacity to deal with phenomena "qualitatively" as well as "quantitatively". A researcher could examine modal choice behavior by simply drawing a cusp catastrophe model and transmitting its properties qualitatively to modal choice behavior, without having to use involved mathematical proofs. For example, one could speculate that there is a "delay mechanism" built in the modal switching process, implying that a relatively large increase in the price of gasoline must be implemented before significant mode switching is to take place. This is a direct qualitative application of the "delay" phenomenon shown in Figures 6.3A and 6.3B. This characteristic is certainly beneficial in policy environments where extensive analysis is prohibitive.

A frequent criticism against C.T., see Sussmann and Zahler in *Behavioral Science* (1978), is that under the structuralist approach the theory merely restates that catastrophic events exist, without accounting for them. Critics argue that the various catastrophic events modeled are not genuine results of the model but simply a part of the assumptions behind it. Specifically, Sussmann and Zahler argue that C.T. models fail to account for a "non-triviality" condition (p. 383) according the which "the mathematical phenomenon must be shown to follow from the assumptions of the model in a non-trivial way and it cannot be just a part of them." Although the argument is involved, put simply it says that one needs to assume initially the existence of discontinuities in order to arrive at a model that would exhibit discontinuities. However, what is and what is not a part of the assumptions in a mathematically stated problem is not itself a trivial question. What may seem to result from deep mathematical reasoning may only be a trick of posterity involving the time to show that the results are a one-to-one correspondence with the assumptions. The choice of calling this correspondence "the results" rather than "part" of the assumptions is simply a matter of preference and tastes.

CATASTROPHE THEORY : A METHOD OF PLANNING

In the previous sections C.T. was presented as a possible technique in planning, with specific references made to certain ad hoc land use, urban economic and transportation examples, as well as to some more ambitious cases. How can C.T. be employed as a method *of* planning is the subject of the discussion to follow. Waters that we are about to explore are uncharted. While many a C.T. analyst has addressed one specific functional example or another no one has attempted to branch into the more general aspects of C.T. as they bear on the essence of planning. As a result, these speculative remarks may prove to be either too analogical, out of place or wrong, especially since C.T. has not yet been tested and validated as a technique *in* planning. However, indications imply that powerful insights may emerge to shed light on events and phenomena which have preoccupied planners' cognition and affected their roles.

Issues of plan relevance, ability to instigate significant changes in the social system, promptness in achieving objectives, capacity to cope with complexity, ability for social learning, and success in the quest for optimality have deeply preoccupied planners. Attempts to address such issues in theory and practice have proven unsatisfactory. This disenchantment has often been though to be due to various constraints imposed upon them in their efforts to perceive and address social problems. C.T., however, provides a new framework to approach these topics.

A social system, at a given moment in its evolutionary stage, expresses through appropriate agents a level of need—a demand function—for action. At the same time the social system through other or identical agents expresses a willingness to undertake—a supply function of—change. The social demand for and the social supply of planning define the social system's "planning potential" defined over the excess demand for or supply of planning. At points where demand equals supply the system is at equilibrium with zero excess demand

(and supply) corresponding to stationary points (points of local optimality) of the planning potential. Such (multiple) points of equilibrium could be either stable or unstable.

The set of all such equilibria points in the space of certain control and behavioral variables defines the *surface of the unfolding of planning*.[14] Control dimensions include the variables affecting the social demand and supply conditions. They may include : available amount of various resources, technology, a social rate of discount, knowledge level, and other exogenous factors or any suitable transformation thereof. Behavioral variables correspond to the level of planned intervention : plan generation, plan evaluation, or plan implementation (action). For any given level of economic resources, knowledge, understanding, etc., there are specific points of zero excess demand (and supply) and a specific equilibrium point on the unfolding of planning. As these control variables change so do the equilibrium points on the planning surface. A specific plan is a particular dynamic path along the surface of the unfolding. Along that surface an infinite number of dynamic paths may be followed, and each path can lead to a specific sequence and type of plan generation, evaluation and implementation processes. The plan's dynamic path is the result of a sequence of values obtained by the control variables (the variables affecting the plan's evolution). A successive set of values of the control variables could be the result of various policies involved in the unfolding of the planning activity. Such policies, for example, could lead planning to a long period of a plan generation phase and then to a catatonic point on the surface. Other policies could lead to a succession of plan generation and plan evaluation phases in a circular manner and never to an implementation phase.

Plan implementation in the form of action by any social agent is a catastrophic event, and so is the succession of a plan's phases—generation, evaluation and implementation. Similarly, the sequence of policy formulation, testing and implementation are catastrophic events. Generally, planned or non-planned actions result in social, political, economic, and ecological transformations, so that the end result of a social action is a point of singularity. Plan generation, evaluation, and implementation are consecutive forms of the planning process, a sequence of morphogenetic events (Thom, 1975), by which the social needs for planning, through a plan, transform themselves from a mere intellectual conception to concrete action. The study of this sequence of transformations requires a theory of plan evolution and its dynamics, a field of research that has remained underdeveloped. Transition to alternative forms, phases in this case, occurs at points of singularities on the surface of the unfolding of planning activity.

The unfolding of the planning surface is a function of the mode of production, consumption and distribution of commodities within a national economy. Such an unfolding is different in capitalist versus socialist states since the process by which social demand for and supply of planning exhibit themselves vary and the number of key factors affecting them differ. Critical points on the unfolding of the surface, the points of singularities, may be interpreted as locations of conflict resolution. Such conflicts are due to the existence of multiple behaviors within the bifurcation set as well as locally conflicting policies which guide the specific path's evolution on the surface of the unfolding.

A plan's transition of phases could be under a "delay" phenomenon, if a planning body prolongs the evaluation phase for as long as possible. Delays occur when a plan enters an area in its unfolding inside the bifurcation set where a number of equilibrium phases could result. Delaying the switching from the experienced phase when entering the bifurcation set for as long as possible is exactly what C.T. refers to as the "Perfect Delay Convention".[15] This delay keeps the system at a point of local optimum, although such optimum may not be the global optimum under the specific values of the control variables. Beyond a certain point inside the bifurcation set the system is better off if it were to switch to an alternative equilibrium state. Bureaucracies are instruments of delay, much as habits are, and there are "perfect delay bureaucracies" and "imperfect delay ones".

Another interesting phenomenon on the unfolding of planning is hysteresis, or the existence of cycles in behavior under various delay conventions. Not being confined to the succession of any particular plan phase, hysteresis is mostly present at the evaluation-implementation and implementation-generation successions.

The catastrophic events mentioned earlier are observed at specific points on the surface of the unfolding of planning when due to local perturbations, *i.e.*, small changes in the values of the control variables. Then sudden switches occur in the system's behavior moving, for example, from plan evaluation to plan implementation. Local perturbations do not necessarily cause catastrophic events, except when they occur at points of singularities on the border of the bifurcation set. Planners and politicians know this phenomenon as "incrementalism", and if occurring simultaneously on many equilibrium points on the unfolding of a surface these perturbations are referred to as "disjointed incrementalism". In planning theory the aspects of "continuity" involved locally through incremental planning are widely documented. C.T. indicates possible discontinuities could occur as well when disjointed incrementalism is the mode of planning and points of singularities are reached. Bureaucracies are local perturbators on the unfolding of the planning activities. C.T. provides a theoretical framework for classifying various policy types : policies that are designed to affect the unfolding are global policies, whereas policies focusing on paths along the surface of the unfolding are local policies.

Local perturbations of the system provide learning of its local properties. Glimpses of the global surface of the unfolding are also possible : the specific part of the whole surface which can be observed will depend on the planner's position on the surface and the history behind. An unexpected jump from one surface of the pleat to another is a genuine addition to the observer's learning of the surface. The issue of locality (the behavior that is of the system at the neighborhood of a point in the control space) is one of the main problems in this area of topology, since it involves the construction of a global topological space out of its local properties. The specific path to be followed on the surface of the unfolding is under the planner's choice of policy. Detective mechanisms, by observing divergence of behavior, delay, hysteresis or jumps, can be used to divert a plan's path on the planning surface if an undesirable point of singularity is imminent. A new, small-scale, local policy can be instituted and move the plan towards a new neighboring path on the surface.

A global policy may have a different effect altogether. It might transform the plan's entire surface to a new surface type thereby causing transition phases to follow different patterns. Such changes are the result of large scale social transformations. These types of unfolding can be more complex when new key policy variables are introduced, or less complex if key policy variables are removed. Planners and policy makers, refer to such large scale interventions as "rational comprehensive planning" when the construction of a specific surface is sought. C.T. provides the elementary structure of all possible unfoldings of all plans. By providing for the most complex phenomena which can occur locally on any plan's surface, it accounts for incremental planning. By providing for a way to classify the surfaces globally it accounts for comprehensive planning. Thus, the elements for a unifying theory of planning is provided within C.T.'s framework.

There may be phases the planning process is undergoing which are stable in the sense that if slightly perturbed they regain their equilibrium status. For example, a plan evaluation phase can be considered as being stable if, slightly disturbed by specific actions from agents expressing either planning demand and/or supply forces, it continues a path involving plan evaluation. However, if slightly perturbed, it jumps to a plan generation or plan implementation phase at a high speed then the initial plan evaluation phase can be said to have been unstable.

C.T. accounts for still another class of policies : those assigned to move the system from a point of disequilibrium to a point of equilibrium ; and those designed to move it from one equilibrium point to another. These motions are related to the concept of stability : only optimal allocations are stable, and stable equilibria are optimal.[16] In planning the concept of stability (intransically related to the concepts of equilibrium and optimality) has not been well developed. The notion is connected to the concepts of "quick" and "slow" functions that play a prominent role in C.T. with the existence of similar functions on the unfolding of planning and/or policy making. According to this distinction, local policies can be of either two types : policies that move the system from a point off the surface of its unfolding (a point of disequilibrium) to a point on the surface of the manifold (a point of equilibrium). These "quick" policies are characterized by relatively expenditious plan generations, evaluation and implementation phases. "Slow" policies are local efforts to move the system from one equilibrium position to another at speeds relatively lower than the speeds encountered in the "quick" case.

CATASTROPHE THEORY : AN EMERGING PARADIGM

Any paradigm's purpose, according to Harris (1979), "is to structure an area of thought and knowledge so as to provide guidance for a number of cognitive activities." C.T. seems to do so. It provides for a construct which can be used in perceiving issues related to the function and essence of the planning process. One can argue that it is richer than existing paradigms since it not only provides for viewing such paradigms as special cases of a general structure, but also for allowing phenomena to be better understood.

C.T., for example, suggests that the conventional way of studying social phenomena has been misconstrued all along. The effort of complete enumeration is simply unnecessary in describing, explaining and predicting social events. We do not need to account for how every factor affects a system's behavior but only look for some key parameters, which may not be recognizable initially. Roles and factors that planners tend to consider relevant might prove to be among those key control variables only in certain instances. Delays in reaching objectives, C.T. suggests, may be an intrinsic aspect of the system to be affected by the plan. Suggestions are also made to the effect that established policies could direct the system along a path quite divergent from expectations, again due to an inherent macrostructure of the system, and its properties.

Finally, C.T. suggests that some portions of a system's behavior may be inaccessible for observation and understanding altogether. They are part of the system's behavior that is unstable. These areas are found in the middle surface of the pleat of the unfolding in the cusp catastrophe of Figure 6.1, and they can only be inferred.[17]

The current state of planning theory leaves many profoundly disconcerted. Problems seem to persist and be ill-defined. At best, social ills do not seem to improve considerably, and at worst they seem to deteriorate. Our basic cognition of problems and our knowledge and roles are incorrectly and partially perceived. Given current paradigms there are some problem aspects that have been kept inaccessible to our search and understanding. Our efforts have been dwarfed by an impression of extreme complexity which we impose upon them and resign ourselves from coping with them or become intellectually neurotic. The avid thirst, exhibited by numerous disenchanted scholars in various social sciences, toward new paradigms indicates the state of affairs in their respective fields.

Is C.T. another paradigm on the unfolding of the surface of scientific planning inquiry ? If so, is it a neighboring one to the existing—or proposed currently ? Our is it a break from them, distant enough, that delays in switching are in operation ? Time will certainly tell.

Fragmentary notions from C.T., as sketched a bove, suggest a new conception of the planning activity. Beyond these general perceptions, however, there are specific, concrete problem areas that can only be understood and approached in a C.T. fromework. In this nation, as well as other western nations, periods of slow economic growth and at times decline are experienced. Critical investment or disinvestment decisions have to be made now and in the future. The effect of such decisions may significantly disrupt and cause specific sectors to collapse. They might affect one nation, but they could be critical in the context of a set of nations. To this author, the need to study and plan for such abrupt changes provides enough justification for the intellectual investment needed to study, develop and apply C.T. for planning and policy making.

FOOTNOTES

1. This paper was originally written in December of 1978. Since then a

number of additional papers appeared in the literature and they are not included here.

2. To the author's knowledge, as of January 1979, these comprise the total body of literature on the applications of C.T. to planning, broader interpreted here as covering urban planning, regional science, transportation planning etc.
3. The narrative to follow will avoid using any mathematical formulae and only few illustrative diagrams will be provided. Mainly, the exposition of the theory here is supplied for pedagogical reasons. For doing this we will draw from the theory's claim that no deep understanding is needed of its theoretical foundations in order to be used, Zeeman (1976). A note of caution is in order : to state the ineffable but in abstract symbols and terms comes at a great cost of precision with a potential for misunderstanding and equivocation.
4. Some researchers content that strictly speaking C.T. is a static theory. In some recent work the dynamic nature of C.T. is exhibited, see Zeeman 1977, p. 81 on the heartbeat and nerve impulse paper.
5. When the governing criterion obtains a local optimum (stationary point) the system is at equilibrium. However, if a local minimum is reached when the criterion is to be maximized, the equilibrium is unstable (and vice versa, if a point of local maximum is obtained when the criterion is to be minimized). The solution is a stable equilibrium if a local minimum is obtained when the governing criterion is minimized, or a maximum is achieved when the criterion is maximized. In order to visualize this, the reader can picture a ball at the bottom of a valley (stable equilibrium), or a ball at the top of a hill (unstable equilibrium).
6. For an excellent discussion of the cusp catastrophe the reader is referred to either Zeeman (1976) or Amson (1975).
7. C.T. demonstrates that the behavior of any system exhibiting catastrophic events can be described by only a few types of basic structures (the elementary catastrophes) no matter what the internal complexity of the system. For more qualitative elaboration of these points the reader is referred to Zeeman (1976), Amson (1974), Flay (1978) and Dendrinos (1978). An ancillary but equally interesting result of C.T. is the assertion that only two elementary structures can describe the catastrophic behavior of any system locally, the fold and the cusp catastrophes.
8. For a deeply behaviorally based, intellectual tradition engrained in Locke and Hume, such statements may sound intellectually profane at worst, pernicious at best and at times solemnly absurd. It can be argued, nonetheless, that such reasoning stems from a Kantian universal structuralism mode of thought.
9. Researchers familiar with mathematical programming know the pains of large-scale system simulation, multidimensional problem solving, multi-criteria decision making, large scale control systems

and other fields of research pursued by social scientists on the premiss that social systems are complex and multi-variable.

10. With regard to the potential function : the complexity of the real governing potential in some instances can be too high so that a researcher might be forced to choose an approximation to it. Such a close and simple approximation to exceedingly difficult potential functions can always be found that it could generate qualitatively the same results. Approximations to forbiddingly difficult energy functions are discussed by C. Benhan and J. Kozak in "Catastrophes in Statistical Biophysics" in *Behavioral Sciences* (1978), p. 357.
11. Berlinski, in *Behavioral Science*, p. 408.
12. The currently prevailing random utility modal choice model in transportation planning is another example. According to this model for any specific value of the control variables for a given individual a unique equilibrium mode is predicted as the chosen mode with a certain probability. C.T. suggests that this may not be the case and for the same values of the control variables a number of modes could be the outcome, depending on the dynamic path the individual trip maker is on (that could be defined by the trip maker's current choice as well as the past ones). For more on this see Dendrinos (1976) where it is shown that the results obtained by the random utility model are special cases of the results obtained by a cusp catastrophe model.
13. Van Parijs (1978) has suggested the use of C.T. for gaining insight into "the central proposition of historical materialism". A cusp catastrophe model is presented on the dynamics of evolution in the mode of the means of production.
14. The surface of the unfolding of the cusp catastrophe is the upper part of Figures 6.1 to 6.4, for simplicity defined as such here.
15. Other delay conventions could also occur—the Maxwell Delay convention and the Imperfect Delay Convention (Amson 1975), are some of them.
16. The notions of optimality and (equilibrium) stability encountered here do not necessarily correspond to the economists' notions. In economics market efficient (equilibrium) solutions may or may not correspond to socially optimal ones, depending upon condition of non-internalization of externalities, existence of public goods and non-constant returns to scale in production or consumption. In C.T. the potential function is intrinsically related to the equilibrium points, i.e., it is not derived independently of the equilibrium solutions.
17. In social systems various "pathologies" would make such unstable states observable : pathological conditions here are defined as the attainment of minima when the maximum of a potential is sought and vice versa.

REFERENCES

Amson, J., 1974, "Equilibrium and catastrophic modes of urban growth, "*London Papers in Regional Science 4. Space-Time in Urban and Regional Models*, Ed. E.L. Cripps (Pion, London) : 108-28.

Amson, J., 1975, "Catastrophe theory : a contribution to the study of urban systems ?", *Environment and Planning B*, Volume 2 : 177-221.

Berlinski, D., 1978, "Catastrophe theory and its applications : a critical review", *Behavioral Science*, Volume 23, No. 5, September : 402-216.

Dendrinos, D.S., 1976, "Two applications of catastrophe theory in transportation planning and urban economics", paper presented at the 22nd Annual Meeting of the Regional Science Association, Toronto, November 1976.

Dendrinos, D.S., 1977, "Sums in capitalist urban settings : some insights from catastrophe theory", forthcoming in *Geographia Polonica*, 1980.

Dendrinos, D.S., 1978, "Urban dynamics and urban cycles", *Environment and Planning* A, Volume 10 : 43-49.

Dendrinos, D.S., 1978, "Operating speeds and volume to capacity ratios : the observed relationship and the fold catastrophe", *Transportation Research*, Volume 12, No. 3 : 191-94.

Dendrinos, D.S., 1979, "Urban dynamics and structural stability : the case of a single city", forthcoming in *Geographical Analysis*, 1980.

Dendrinos, D.S., H. Mullally, 1979, "Fast and Slow Equations : the development patterns of urban settings", paper presented at the Annual Meeting of the Regional Science Association, Los Angeles, November.

Flay, B.R., 1978, "Catastrophe theory in social psychology : some application to attitudes and social behavior", *Behavioral Science*, Volume 23, No. 5: 335-350.

Harris, B. ; Wilson, A., 1978, "Equilibrium value and dynamics of attractiveness terms in production-constrained spatial-interaction models", *Environment and Planning A*, Volume 10, No. 4 : 371-388.

Harris, B., 1979, *A Paradigm of Planning*, University of California Press, forthcoming, 1980.

Isard, W. ; Liossatos, P., 1977, "Models of transition processes", *Papers of the Regional Science Association*, Volume 39, 27-60.

Mees, A., 1975, "The revival of cities in medieval Europe", *Journal of Regional Science and Urban Economics*, Volume 5, No. 4, December : 402-426.

Poston, T. and Stewart, I., 1978, *Catastrophe Theory and Its Applications*, Fergson—Pitman, Belmont, California.

Poston, T. ; Wilson, A., 1977, "Facility size versus distance travelled : urban services and the fold catastrophes, *Environment and Planning A*, Volume 9 : 681-686.

Sussmann, H. and Zahler, R., 1978, "A critique of applied catastrophe theory in the behavioral science", *Behavioral Science*, Volume 23, No. 5 : 383-389.

Thom, R., 1975, *Structural stability and morphogenesis : an outline of a general theory of models*, W.A. Benjamin, Inc., Reading, Massachusetts. Originally published in 1972 as *Stabilite structurelle et morphogenese. Essai d'une theorie generale des modeles.*

Wilson, A.G., 1976, "Catastrophe theory and urban modelling : an appilcation to modal choice", *Environment and Planning A*, Volume 8 : 351-356.

Woodcock, A.E.R. and Poston, T., 1974, "A geometrical study of the elementary catastrophes", *Lecture notes in Mathematics*, 373, Spinger, Berlin.

Van Parijs, P., 1978, "Theorie des catastrophes et materialisme historique", *Revue Francaise de Sociologie*, Vol. XIX, No. 2, Avril-Juin : 195-220.

SEVEN

M.J. OXLEY

ECONOMIC THEORY AND URBAN PLANNING

WITHIN any process, contradictions can be more easily recognised, if not corrected, when there is some common force linking the composite elements. Logic would suggest unity of purpose regarding the goals or net outcome of the forces. However, when the process is town planning it is the very question of purpose which tears forces apart. An interested 'outsider' could gain nothing but disuasive confusion by looking to planners for a definition of their goals, and quite possibly anger at realising the nonchalance with which public actions, without ordered public purpose, are undertaken.

What are urban planners trying to do ? Are they aiming 'to resolve conflicting uses to which land is put' ? By what criteria do the uses conflict ? 'To maintain a good balance between work and living places' ? What is this balance ? 'To improve the quality of the environment' ? According to which standards ? 'To provide more and better quality housing' ? Why is this necessary ?

For all that has been written on the subject, planners have no framework within which to clarify their ends, to ascertain why these should be their ends, or to fit means to ends. In short, urban planning is sadly lacking in a theoretical foundation. The current vogue of interdisciplinary approaches has done nothing to foster the development of such a foundation

Reprinted from *Environment and Planning* A, 1975, Vol. 7, pp. 497-503.

for, while economics, scciology, geography, architecture, Old Uncle Tom Aesthetics and all are alleged to vaguely have 'something of value' to contribute, there is no agreement on a common framework within which these individual contributions are to be made. The result is a ragbag of ill-defined goals and means, with potential conflict between and within both.

Extreme wooliness and a lack of a systematic approach to the subject of goal and policy formulation is apparent from the literature addressed to this issue. There is a concern for order without defining that order. Foley (1960) distinguishes three ideologies :

1. "Town Planning's main task is to reconcile competing claims for the use of limited land so as to provide a consistent, balanced, and orderly arrangement of land uses."
2. "Town planning's central function is to provide a good (or better) physical environment ; a physical environment of such good quality is essential for the promotion of a healthy and civilised life."
3. "Town planning, as part of a broader social programme, is responsible for providing the physical basis for better urban community life."

The theme prevails, often implicitly, of environmental determinism (see Reade, 1968). In its most severe form this disease becomes architectural determinism (see Broady, 1968). Dennis (1970) identifies 'Usherism'—the belief that bricks and mortar are themselves of significance. The prime role of alternative specialisms have been variously argued, whether it be the supremacy of the architect or of the location theorist (McLoughlin, 1970, Chapter 3). Stamp (1963, page 10) notes "the attempt to improve living conditions, which is the object of town and country planning. Such is, indeed, the field of applied geography", and Freeman (1958, page 13) claims that "planning has an inescapable geographic bias".

The question arises of the compatibility of these conclusions with such views as "the comprehensive ideal of planning and thus its desire for comprehensive goals, is founded on the belief that private decision markers with a particular mission . . . and public decision makers with a particular mission . . . are concerned primarily with their own family or business or part of the public service. The planner seeks to coordinate integrate design and decide on the basis of the *total-public welfare overall* [my italics] (or most) aspects of the life of the city or region, and with a longer time-span in view" (McLoughlin, 1970, page 107). For this comprehensive approach there seems to be no organisational thought-structure geared to identifying aims and alternative courses of action, despite the work of, for instance, McLoughlin (1970) and Chadwick (1971). If there was, as Cullingworth (1970) claims, an early "appreciation of the *necessity* for interfering with market forces and private property rights in the interests of well being" it is astonishing that the appreciation has not led to a reinterpretation within the mainstream of planning methodology, of the body of theory relevant to these very issues. This theory is within welfare economics. In what circumstances does this *necessity* arise and by what means may the need be met ? This is the central concern of planning and of welfare economics. It is the *central concern of no other* discipline. The parallel has only occasionally been *directly* recognised and rarely fully exploited. From the mixing-pot of statements as to planning's purpose the odour that lingers longest surrounds those concerned with reaching a position which is 'best' for society. This has a direct analogy with welfare optimisation—an analogy which, if understood more clearly by planners and developed more fully in terms of operational criteria by economists, could conceivably form the stock of a body of planning methodology onto which other subjects might graft their relevant expertise. A grand objective, may be, but this has to some extent occurred in the development of cost-benefit analysis although, even here, the welfare implications and conflicts of the technique with welfare propositions are rarely pressed home in planning literature. These failures are in part a function of the complexity of the expositions of welfare economics. Let us attempt an elementary approach.

What does welfare optimisation mean?

Three elements of the optimum performance of an economic system may be identified (Winch, 1971, page 77) :

1. Efficiency in production.
2. Efficiency in exchange.
3. Equity in distribution.

'1' and '2' together comprise the efficiency criterion or the criterion for Pareto optimality, defined as a situation from which no reallocations of scarce resources can improve the welfare of some without reducing the welfare of others. (For a fuller exposition of this see, for example, Bator, 1957.) '3' requires that income and wealth be 'fairly' distributed. There is no suggestion that such 'optimality' is without value judgement, but the value judgements can be made explicit.

Within given constraints, an efficient allocation can arguably be achieved under a system of perfect competition, the conditions for optimality thus being best promoted by a total absence of planning (for further details see the appendix). If the 'best system' is one in which there is no planning, why, in economic terms, is there any need for this activity ? A naivety put about by planners, and others, is that economics and economists believe that a system of market forces is the preferred state and that any intervention rationale must be based on, for example, sociological or political reasoning. In fact economics itself provides for the grand destruction of noninterventionist arguments.

Market failure

The conclusion that perfect competition can promote optimality has strong expression in the 'Invisible Hand Doctrine' of Adam Smith and forms part of the argument for free enterprise and against 'intervention'. The holes in the 'no intervention is best' argument are many and large, and are well documented. (For a broad summary see Bator, 1958.) For example :

(1) Perfect competition is a theoretical market structure with such restrictive assumptions that is fails to accord with reality (see any basic economics text book—for example Samuelson, 1973, or Lipsey, 1973). Perfect knowledge, homogeneous product, and free entry and exit to markets are real-world rarities. This has been turned to provide an argument nevertheless, for using the model as a basis for comparison and has thus itself provided the basis of certain arguments for intervention—intervention designed to remove restrictive barriers and promote competition. However, the system does not cater for nontraded costs and benefits. The Theory of the Second Best (Lipsey and Lancaster, 1956) provides a second major challenge to this approach by showing that achieving efficiency-conditions in some but not all markets may well move away from optimality rather than towards it.

(2) It is the equilibrium conditions of perfect competition that are compatible with the Pareto efficiency criterion. Without equilibrium we can make no such judgement. Disequilibrium is the rule rather than the exception. The weakness here stems from the methodology of comparative statics within which the theory has expression. How do we tackle problems of dynamics ? What of the costs and benefits of adjustment through time ? What can be the policy-significance of the theoretically inconsistent concept of equilibrium ? "The notion of equilibrium – whether in the market for peanuts or in the level of employment in a private enterprise economy—is part of the wishful thinking of neoclassical economists, which betrayed itself, in the large, in their inability to understand the great slump and, in the small, in judging the arguments about supply and demand" (Robinson and Eatwell, 1973, p. 164).

(3) Perfect competition may be unable to sustain Pareto optimality given increasing cost industries where, by definition, marginal cost exceeds average cost. Winch distinguishes two cases :

(*a*) Where all factors are privately owned and at least one cost curve is upward sloping, which is of significance from a distributional rather than an efficiency viewpoint.

(*b*) Where one factor is not privately owned and is treated as a free good, which is of more significance, including efficiency considerations, for urban problems. "Many such cases have arisen and our society appears to be remarkably slow in learning the lesson common to all. Over-grazing of medieval common (free) pasture resulted in poor quality cattle. That lesson was learned centuries ago but today we still overcrowd our (free) roads with vehicles and over-pollute our (free) air and water" (Winch, 1971, p. 103).

(4) Perfect competition is incompatible with decreasing-cost industries. This might be due to increasing returns resulting from indivisibilities. This is the problem of scale. In such cases long-run average cost is above long-run marginal cost, and charging a price equal to marginal cost would result in continuous losses. Urban transport, water supply, and sewerage provision may fall within this category.

(5) Perfect competition does not cater for the existence of interdependencies in production and utility functions. When nontraded, these interdependencies raise the issue of externalities. The existence of an external effect requires at least two bodies, and a rough and ready classification system acknowledges this fact. Thus we may have :

Producer-producer externalities. An example with reference to river pollution : Firm *x* sets up a business upstream from firm *y* which had previously used the relatively clear water for washing its product. Now firm *x* dumps waste matter from its production process into the river, so imposing on *y* the added cost of water purification if use of river water in its production process is to continue.

Producer-consumer externalities. The classic example is the factory chimney (Pigou, 1932) which pours out black smoke onto neighbouring washing lines so increasing laundry bills. Social costs are thus imposed. Pollution and congestion problems whether they relate to water, air, or road-use are analytically of a similar nature—firms imposing, in the absence of intervention, nontraded costs.

Consumer-consumer externalities. If one allows for the (intuitively realistic) possibility that an individual's utility is a function not only of his own consumption of goods and services but of others' consumption also, 'bandwagon' and 'snob' effects provide further cases of nonmarket interdependence which prevent the achievement of Pareto optimality in a process of perfect competition.

The existence of external effects drives a hard wedge between private and social costs. Thus market operations based on individual decisions do not maximise social objectives. Here is the mythology of the notion that each man seeking to maximise his own good will automatically maximise that of society—the fallacy of Benthamite additive philosophy.

(6) The market mechanism fails to provide a criterion for the provision of public goods (Samuelson, 1954), those for which the consumption by one individual does not reduce that available for others, and suppliers find it difficult to exclude nonpayers. Given this situation, consumers will have no incentive to reveal their preferences for such goods for they receive the benefit whether they pay or not. A television or radio signal provides one of the few examples of a near 'pure' public good, but quasipublic goods containing an element of 'publicness' abound in the urban economy—parks, open spaces, roads (up to the point where they become congested), police protection, etc.

(7) The costs of operating a market may be so high that alternative allocation mechanisms save resources. This may be true not only for cases where publicness or externality arises but for any situation in which property rights are poorly defined or vague. Markets in road and bridge-usage as well as charges for entry to public parks are, we know, feasible but are not always the preferred way of allocating resources if collection costs are high or if the price charged pushes consumption away significantly and leaves underutilised resources.

(8) Even the most extreme advocates of market freedom rarely make claims in terms of the equity aspect of the price system. If society is unwilling to accept the arbitrary nature of the distribution of rewards and satisfaction provided without intervention, indeed if one sees this income distribution as the

very root of many urban problems, the market system has clearly failed to promote optimality.

It might be decided that a more fundamental challenge is to be posed not only to marketism but to the welfare theory that has been sketched here (see, for example, Gintis, 1972). What if society expresses preferences not only in the derived utility space but also in input space such as to argue for the redistribution of the initial endowment of resources ? Implementation will require reallocation and redefinition of the property rights which are traded in markets. What price a house in a society where there are no penalties for one forcing occupants out of dwellings at gunpoint and then assuming ownership ? If stealing and appropriation by means other than exchange for money are acceptable, market operations clearly cease to have significant value. One buys the right to usage and sale—as for any commodity. The definition of property rights may be seen to require, in itself, intervention. We may point to the paradox that the very operation of markets is secondary to 'intervention' to determine the property rights there traded.

A market system, it can thus be argued, works well in terms of both efficiency and equity criteria only when property rights are well defined ; the initial distribution of property rights is acceptable ; factors are mobile ; information and transaction costs are negligible ; externalities in production, consumption, and exchange are absent ; public goods are absent and the distribution of rewards is acceptable.

Without these conditions satisfied in all markets, the 'Theory of the Second Best' warns of the danger of attempting optimality by striving for their achievement in some but not all markets. The optimality characteristics of perfect competition are those of equilibrium. The real world is one of adjustment to equilibrium. Typically, urban problems are adjustment problems. The market system offers little in the way of comfort to those paying the costs of adjustment.

Faith in the free working of the price mechanism has arisen out of either an ignorance of, or lack of concern with, the nature of market failure and a misunderstanding of the nature of Pareto optimality. The planner is, by his existence, challenging either the efficiency or equity aspect of the market system or both. In his challenge he would be wise to show an awareness of the nature of market failure, and indeed within the framework of this breakdown formulate his goals and alternative courses of action to achieve those goals.

The nature of the challenge will vary with the point at which one enters the economic system and will be subject to political and social considerations. The tradition of economic science is to render these as value judgements and thus, by definition, beyond its scope. As the myth of value-free social science is increasingly revealed, the planners and economists together might be well advised to evaluate systematically the implication of alternative social and political viewpoints. The fundamental issue is the level at which the challenge to marketism is aimed. Is it in terms of the inputs to the market system, its internal operations, or its outputs ?

The first-level challenge, to question the inputs, is perhaps the most radical stance of all. Do we accept existing ownership rights regarding land and property or do we wish our intervention to start by redefining and redistributing property rights ? To acquiesce in the acceptance that existing laws and facts of inheritance decree inalienable rights of ownership, for which any encroachment or confiscation must be compensated, is to declare forever the supremacy of the *status quo*, to mix a permanent recipe for conservatism. The property rights over land and their consequence for land use are clearly at the heart of the urban planning process. In this context the planning machine, if viewed as an allocative machanism alternative to the market, must inevitably be considered as passive rather than innovative in its 'yes'/'no' approach to planning permissions. A more positive role may not be possible without basic changes in the existing rights of ownership going beyond compulsory purchase orders. The various forms of community ownership and land nationalisation that have been suggested are likely to be accepted or rejected on the basis of uninformed prejudice or political expediency. This decision should not be determined in isolation by politicians and a few academic theorists and then become a *fait accompli* for planners, but rather form part of the planning process, one which starts with active concern amongst planners to research the likely outcomes of the various proposals. Thus active ongoing research into mooted political changes becomes necessary as an integral part of planning. We can question the proportion of research, currently being undertaken by the various research institutions and university departments claiming to specialise in the areas of urban studies or town planning, that is inquisitively policy orientated. If it is more than a tiny minority, one can only assume the process of dissemination to be painfully slow. By the time the one millionth esoteric mathematical simulation is formulated, an instant self-destruct device will have been built-in to justify its existence in the face of the political actions that have occurred, quite independently, during its gestation.

While any action at this first-level challenge is likely to involve politics of the extreme left or right, more moderate activities appealing to the political centre may be applicable at the second-level challenge, the market's internal operations. Here the antimarket claims can be couched in terms of the objections raised in the previous 'Market failure' section. The economics of externalities might usefully become not only part of planning theory but a base for empirical investigations into the viability of alternative intervention techniques.

In transport planning are planners not facing the problems of externalities in the use of road space ? Should not planning be aimed at correcting the externality ? The methods are many and go far beyond building motorway boxes or road-pricing into the area of redefining property rights with respect to road use, and redistributing the costs and the benefits of road usage by alternative transport-pricing and ownership systems. Is the problem in part one of *individual* ownership of cars and *individual* decision making as to the nature and frequency of their use ?

In land-use control, why does the planner want a say in whether plot A next to plot B (a house) is used as a brewery or a supermarket ? Is it because housing and brewing are incompatible uses, because there is a 'need' for a supermarket in the area, or because the brewery will add to traffic congestion, or all three ? Each provides examples of market failure and to each problem the gamut of solutions runs far beyond the physical determinism of land-use control and, again, into the areas of property-right definition, taxation, subsidy, compensation, and income-redistribution type approaches.

Arguments about reform of the rating system take place in an unnecessary vacuum between 'practical changes' based on political bigotism or mere inclination, on the one hand, and a body of market failure theory on the other. The need for rates (if indeed there be one) may be seen to arise from the failure of the market to cope with, inter alia, the provision of public goods, the treatment of externalities, and the realisation of economies of scale which the local authority attempts and for which it accordingly requires finance. We need to know more, however, about the economics of the various methods of financing and allocating public open spaces, sewers, and road space for example. Individually and from many uncollated sources we have scraps of information but, again, a full body of tested alternatives applicable to policy issues is lamentably absent.

The third-level challenge is aimed at the outcomes of marketism. The inputs to, and the workings of, markets are basically accepted, but the distribution of satisfactions or incomes that result is not. Action is then aimed directly at redistributing incomes either at the national or the local level. Many so called 'town planning issues' might become radically redefined as objects for basic changes in the taxation or subsidy systems. Housing is but one area where such an approach would be significant. If the 'housing problem' is reduced to its fundamental components, one might view it essentially as a problem of certain individuals or groups in society being unable to bid in the market for 'decent' accommodation because of insufficient income, a basic answer in the form of income redistribution is revealed. That poverty of income is the basis of the problem, in the face of which rent control and other methods of intervening in the working of the housing market is irrelevant, is argued cogently by Pennance (1968).

Of planning as an essentially nonmarket distributive process we know very little.

Implications of a welfare-economics approach

Within a market-failure approach, the thoughts of an urban planner might be organised in terms of :

1. What form or forms of market failure underpin my objective and at which level—inputs, working mechanisms, or outputs—am I challenging the market system ?
2. What forms of intervention exist and have been suggested in the economic literature to deal with this form of failure ? (See for example the readings and associated references in Layard, 1972.)
3. Which method is applicable in this particular case ?

Such an ideal state remains a topic for naive speculation without a number of practical prerequisites :

(*a*) The education of planners in the economics of market failure - an education which would stress the value of an economic framework within which actions may be contemplated. The enormous size of this task is apparent from reviewing the particularly inadequate recognition of economic analysis given in planning textbooks. For example, in his *An Introduction to Town and Country Planning*, Ratcliff (1974) seems to fail totally to recognise the significance of welfare in economics. Instead he notes that "Economics...places greater emphasis upon the maximisation of efficiency and production whereas town planning seeks to maximise the welfare of the community." There follows a list of seven areas in which "The economist, with his tools of analysis, knowledge of markets and understanding of location can assist the process of town planning", which gives no mention of the potential role of economics in determining the nature and means of market intervention.

(*b*) The evolution of a body of policy-orientated research founded on a 'market failure' framework. Land nationalisation, changes in the rating system, and new housing mortgage arrangements are but three areas in which such investigation is essential.

(*c*) An attempt at an education of the general public in the rationale of planning within our elementary market-failure context. Arguments about the impossibility of the task should not be readily accepted from an elite who attempt a retention of knowledge for the few. If the means by which economics, and welfare economics in particular, is explained, interpreted, and its ideas exchanged and disseminated is a foreign language, or available in limited editions only, misunderstandings about how the economic system operates and the ways in which the results of the system can be modified, leaves some in blissful ignorance but for others creates misplaced antagonism and fruitless attempts to alter the system. Informed public discussion on the purpose of planning is an essential prerequisite to a consideration of the more high-minded hope of 'public participation'.

(*d*) In evaluating policy options suggested by a welfare-economics approach, consideration is necessary of the political and philosophical stances with which each is compatible. A consequence may be the sound involvement of urban planners in the wider debate of capitalism versus collectivism, and the future of the mixed economy.

The more all this seems rather trite to economists but strange to planners, the wider is the void thus revealed.

CONCLUSION

As planning seeks to provide alternative allocations to those of the market system its rationale might start from :

(1) the unacceptability of the distribution of factor ownership ;
(2) the unacceptability of the efficiency of a market system ; or
(3) the unacceptability of the distribution of satisfactions resulting from the system.

Any or all of these starting points will provide a theoretical basis for intervention, and suggest a range and radicality of solutions—some outside of the present institutional framework.

Economic theory has the advantage of identifying and ordering the problems within such a framework and providing a theoretical basis for the solutions. Its weakness is in fully transforming its propositions into operational objectives. Until planners turn themselves to the theoretical problems of economics, and economists to the practical problems of planning, we must be content with planners who act as if to cure measles by painting over the spots, and economists who provide paint which no brush can apply.

REFERENCES

Bator, F.M., 1957, "The simple analytics of welfare maximisation", *The American Economic Review*, March.

Bator, F.M., 1958, "The anatomy of market failure", *Quarterly Journal of Economics*, August.

Broady, M., 1968, *Planning for People : Essays on the Social Context of Planning* (Bedford Square Press of the National Council of Social Services, London).

Chadwick, G.F., 1971, *A Systems View of Planning—Towards a Theory of the Urban and Regional Planning Progress* (Pergamon Press, Oxford).

Cullingworth, J.B., 1970, *Town and Country Planning in England and Wales* (Allen and Unwin, (London).

Dennis, N., 1970, *People and Planning : The Sociology of Housing in Sunderland* (Faber and Faber, London).

Foley, D.L., 1960, "British town planning : one ideology or three", *British Journal of Sociology*, 11 (3), 211-231.

Freeman, T.W., 1958, *Geography and Planning* (Hutchinson Library, London).

Gintis, H., 1972, "A radical analysis of welfare economics and individual development", *Quarterly Journal of Economics*, 91 (4).

Layard R. (Ed.), 1972, *Cost-Benefit Analysis* (Penguin Books, Harmondsworth).

Lipsey, R.G., 1973, *An Introduction of Positive Economics*, 3rd edition (Weidenfeld and Nicolson, London).

Lipsey, R G., Lancaster, R.K., 1956, "The General Theory of the Second Best", *Review of Economic Studies*, 24, number 63.

McLoughlin, J.B., 1970, *Urban and Regional Planning—A Systems Approach*, Chapter 3 (Faber and Faber, London).

Pennance, F.G., 1968, "Housing market analysis", Hobart Papers (Institute of Economic Affairs, London).

Pigou, A.C., 1932, *The Economics of Welfare*, 4th edition (Macmillan, London).

Ratcliffe, J., 1974, *In Introduction to Town and Country Planning* (Hutchinson Press, London).

Reade, E., 1968, "Some notes towards a sociology of planning : the case for self awareness", *Journal of the Town Planning Institute*, 54 (5), 214-218.

Robinson, J., Eatwell, J., 1973, *An Introduction of Modern Economics* (McGraw-Hill, London).

Samuelson, P.A., 1954, "The pure theory of public expenditure", *Review of Economics and Statistics*, 41, 387-389.

Samuelson, P.A., 1973, *Economics*, 9th edition (McGraw-Hill, London).

Smith, A., 1776, *Wealth of Nations* (Penguin edition, Harmondsworth, 1970).

Stamp, L.D., 1963, *Applied Geography* (Penguin Books, Harmondsworth).

Winch, D.M., 1971, *Analytical Welfare Economics* (Penguin Books, Harmondsworth).

APPENDIX

OPTIMISATION CONDITIONS

The economic system is viewed as one of inputs of factors of production and outputs of goods and services from which utilities are derived. Only private goods and services exist, utility functions are independent and all production functions convex. Thus all goods and services are traded in markets ; satisfactions derived from one commodity do not influence those derived from other commodities and there are constant returns to scale. The conditions for maximum efficiency to be achieved in such a system may be explained diagrammatically in terms of a two factor (*l* and *k*), two product (*A* and *B*), two person (I and II) world, but generalisation to a multifactor/multiproduct/multi-person world is possible (Bator, 1957).

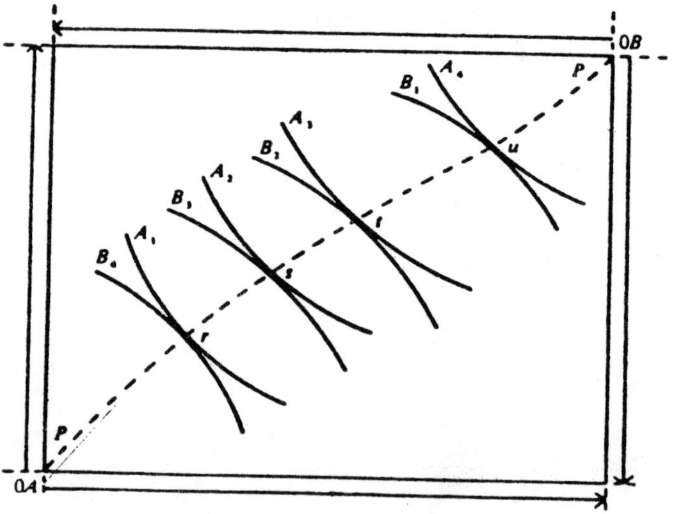

FIG. 7.1 : *Input space*

In Fig. 7.1 the horizontals of the box (an Edgeworth-Bowley production-contract box) indicate the amount of *k* available, the verticals the amount of l. With OA as the origin, A_1 plots the alternative combinations of l and k that can produce a given physical volume of A. A_2, A_3, A_4, plot these combinations for increasing amounts of A. With OB as the origin B_1, B_2, B_3, B_4, perform similar functions for product B. A_1, A_2, A_3, A_4, B_1, B_2, B_3, B_4 are isoquants. The slope of an isoquant gives the rate at which one factor may be substituted for another in the production of the product (the marginal rate of transformation of l for k). $R^A_{T(lk)}$ is the slope giving the marginal rate of transformation of labour for capital in production of good A. $R^B_{T(lk)}$ is the slope giving the marginal rate of transformation of labour for capital in production of good B. Only when

$$R^A_{T(lk)} = R^B_{T(lk)}\,, \qquad \text{(A.1)}$$

can no reallocation of l and k between A and B lead to a greater production of one without an accompanying reduction in the output of the other.

Thus only such combinations as identified by r, s, t, u are 'input efficient'. The line PP is the locus of the points of isoquant tangency where equation (A.1) holds. This identity is the first necessary condition for Pareto optimality.

To derive the second condition we move from the input space of Fig. 7.1 to the output space of Fig. 7.2, where the horizontal axis plots outputs of A, and the vertical axis outputs of B. Combination r (A_1 and B_4) is reidentified as R, and s as S. PP (a production-exchange contract curve) becomes P_1P_1 (a production-possibilities curve), which shows all the efficient outputs of A and B possible from l and k with a given technology. The opportunity cost of A in terms of B can be measured by the slope of P_1P_1 which equals the marginal rate of transformation of A into B ($R_{T(AB)}$).

So far the concern has been only with efficiency in production (points from which no more A can be produced without losing some B). What of efficiency in exchange (points from which no redistribution of A and B between I and II can improve satisfaction) ? The problem is the allocation of a given output (say S) between consumers. Superimposed on the output space of Fig. 7.2 is an Edgeworth-Bowley exchange-contract

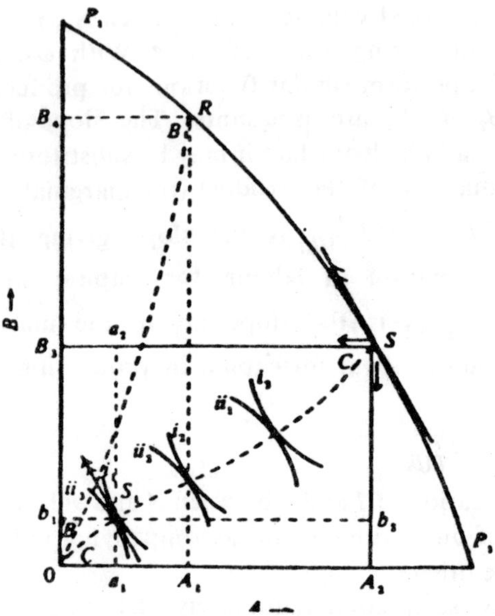

FIG. 7.2 : *Output space*

box OB_3SA_2. With O as the origin, i_1 shows alternative combinations of A and B to which consumer I would be 'indifferent'. i_2, i_3, show this for higher levels of satisfaction. With S as the origin, ii_1, ii_2, ii_3, perform this function for consumer II. i_1, i_2, i_3, ii_1, ii_2, ii_3 are 'indifference curves'. The slope of an indifference curve shows the consumer's marginal rate of substitution (R_s) of one product for another. In Figure 7.2, only at points of indifference-curve tangency can there be no increase in the utility of I or II without an accompanying reduction in that of the other. At tangency the marginal rate of substitution between products A and B for consumer I is equal to that for consumer II, that is :

$$R^{I}_{S(AB)} = R^{II}_{S(AB)}. \qquad \text{(A.2)}$$

The exchange-contract curve (CC) is the locus of all such tangencies. For output S, all points on CC are superior to all points off CC in that they are 'exchange efficient'. Other ex-

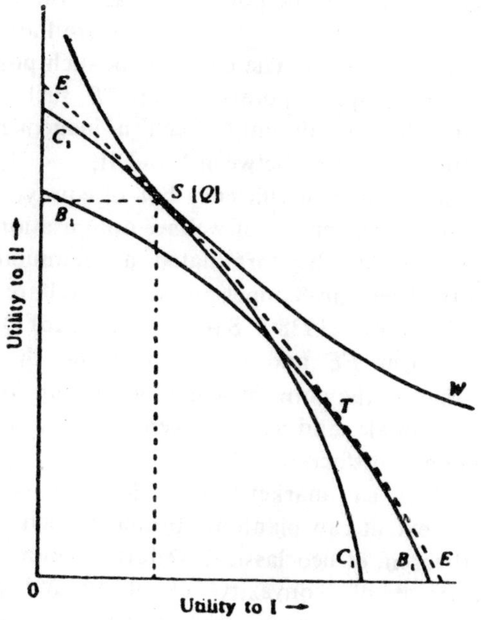

FIG. 7.3 : *Utility space*

change-contract curves may be traced for other outputs (for example, *BB* for *R*).

Pareto efficiency, requiring that equations (A.1) and (A.2) are both satisfied, is achieved at a combination such as output *S* and exchange point S_1. Here

$$R^{A}_{T(lk)} = R^{B}_{T(lk)} = R^{I}_{S(AB)} = R^{II}_{S(AB)}.$$

Generalising to a multifactor/multiproduct/multiperson world,

$$R_T = R_S \qquad \text{(A.3)}$$

(where R_T is a common marginal rate of transformation, and R_S a common marginal rate of substitution). With A_1 produced, a_1 is allocated to I and a_2 to II, and from output B_3, b_1 is allocated to I and b_2 to II. An infinite number of points of Pareto efficiency can be identified in Fig. 7.2.

Figure 7.3 shows *CC*, relating to output level *S*, redrawn in utility space as utility possibility curve (UPC) C_1C_1. Other UPCs for other output levels may be added, such as B_1B_1 for output

R. On each UPC there will be points such as S on C_1C_1 (equivalent to S_1 on CC) and T on B_1B_1 which are Pareto efficient in production and exchange. The envelope of such points is EE. But how can one compare points along EE such as S and T all of which are Pareto efficient ? Such a judgement involves comparisons in utility space between I and II.

This is a question not of efficiency but of equity. So, resort is made to the third element of welfare optimisation. If distributional judgements can be formulated a community indifference curve WW (Bergson-Samuelson social welfare function) can be drawn (Bergson, 1938 ; Samuelson, 1956) Q, the point of tangency between EE and WW, now defines the 'Optimum Optimorum', that is, the point at which, given our assumptions about inputs, rationale, and value judgements, society is at its point of optimum satisfaction.

To what extent can market forces achieve a position such as Q ? To what extent can planning promote such a position ? A standard theorem of neoclassical welfare economics is that, given the constraints of 'convexity' of utility and production functions, one market structure—that of perfect competition—is unique is that it can sustain the efficiency conditions of Pareto optimality.

The optimality of perfect competition ?

Given perfect competition, then a common set of prices for each factor and product prevails throughout the economy, in which individual producers and consumers are unable to influence these prices.

Producers attempting to maximise profits in the face of common prices will try to minimise costs by equating the marginal rate of technical transformation (R_T) with the factor price ratio, thus organising production such that :

$$R^A_{T(lk)} = \frac{p^k}{p^l},$$

where

p^k is the price of capital,

p^l is the price of labour, and

$$R^{B}_{T(lk)} = \frac{P^{k}}{P^{l}},$$

therefore

$$R^{A}_{T(lk)} = R^{B}_{T(lk)}$$

which is equivalent to (A.1) above. (A.4)

The consumer, attempting to maximise the return from a limited income, will allocate expenditure such that :

$$R^{I}_{S(AB)} = \frac{P^{A}}{P^{B}}$$

where

P^{A} is the price of A,

P^{B} is the price of B.

Given common prices, this ratio is the same for all consumers. Thus in a two-product/two-person case,

$$R^{I}_{S(AB)} = R^{II}_{S(AB)} = \frac{P^{A}}{P^{B}} \qquad \text{(A.5)}$$

which is equivalent to (A.2) above.

Profit maximisation requires that $P^{A} = c^{A}$, and $P^{B} = c^{B}$ where

c^{A} is the marginal cost of A,

c^{B} is the marginal cost of B.

Thus

$$\frac{c^{A}}{c^{B}} = \frac{P^{A}}{P},$$

and

$$R_{T(AB)} = \frac{c^{A}}{c^{B}} = \frac{P^{A}}{P^{B}},$$

therefore,

$$R_{T(AB)} = R_{S(AB)}.$$

Generalising to a multifactor/multiproduct/multiperson world

$$R_{T} = R_{S}, \qquad \text{(A.6)}$$

which is equivalent to (A.3) above.

REFERENCES

Bator, F.M., 1957, "The simple analytics of welfare maximisation", *The American Economic Review*, March.

Bergson, A., 1938, "A reformulation of certain aspects of welfare economics", *Quarterly Journal of Economics*, 52, 310-334.

Samuelson, P.A., 1956, "Social indifference curves", *Quarterly Journal of Economics*, 70, 1-22.

EIGHT

ERIC L. HYMA

THE MULTIPLE—OBJECTIVE CONCEPTUAL FRAMEWORK FOR ENVIRONMENTAL QUALITY ANALYSIS

THE MULTIPLE-OBJECTIVES PLANNING PROCESS

A RATIONAL planning process adapted to address multiple objectives provides an excellent conceptual framework for environmental quality analysis (Hufschmidt, 1971). A major obstacle to this integration is the overwhelming amount of information on aspects of environmental quality. Some method must be employed to collapse the information so that it can be utilized in policy making.

The rational planning model (Banfield, 1955 ; Janis and Mann, 1977) consists of five basic steps : (1) identification of a problem or opportunity, (2) setting objectives, (3) development of planning guides and criteria, (4) formulation of plans, and (5) selection and evaluation of plans (Fig. 8.1). These basic steps have been refined to include design and testing of alternatives, data collection, prediction, simulation, explicit questioning of assumptions and sensitivity of outcomes, and re-examination of alternatives, with various alternative feedback loops (Quade, 1966). Recent versions of the rational planning model emphasize the importance of implementation.

Decision makers need summary information on the identification, prediction, and evaluation of environmental quality.

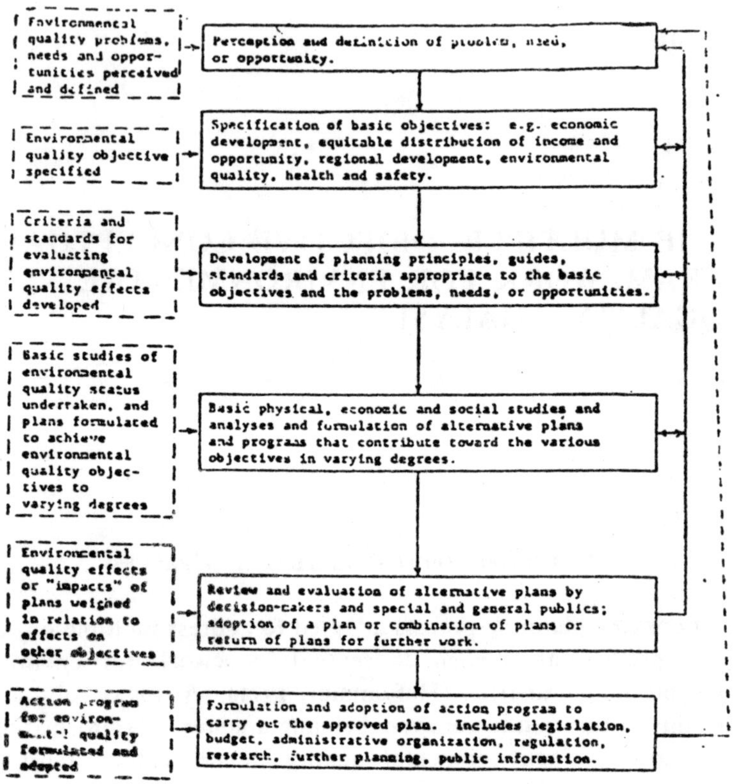

FIG. 8.1 : *A Water Resource Planning and Decision Process with Environmental Quality Relationships*

Identification includes description of the existing environmental system and awareness of problems and opportunities. Prediction is the analysis of expected changes in the system whether exogenous, natural changes or impacts of human activities. Evaluation places positive or negative values on environmental changes and compares tradeoffs associated with various alternatives. Raw data are too cumbersome to use effectively. On the other hand, summary indicators may contain unrealistic assumptions.

Environmental quality indicators and methods are useful in making resource allocation and regulatory decisions, setting priorities, and studying changes in the state of the environment. The information from indicators and methodologies can be used in the development of objective functions, decision criteria, weights, and performance measures, key components of the multiple-objective planning process.

Interest in multiple-objective planning grew strong in the mid-1960s because of increased dissatisfaction with the inherent incompleteness of gross national product and other measures of economic welfare. The association between economic growth and social well-being was challenged (Mishan, 1969). There was also a reaction against the dominance of purely monetary variables in the decision-making calculus (Campbell, Converse and Rodgers, 1976).

Economic indicators focus on only one aspect of the social welfare function. The quality of life involves other social, political, psychological and environmental dimensions, many of which are qualitative or subjective variables (Stiftel and Hyman, 1981). "Quality of life" is an ambiguous phrase which has yet to be pinned down exactly on the individual or societal levels and perhaps is never will be defined rigorously. Nevertheless, it is a concept of which everyone has some definite impression, and it can be a useful way of looking at the desired endproduct of multiple-objectives decision making.

At any rate, many non-conomic elements are recognized as important by consensus. Often, these objective conditions can be quantified to some extent in physical terms or we can determine the subjective preference functions of individuals and groups for various scenarios.

Some skepticism is warranted because the phrase "quality of life" is nebulous and means different things to different people ; thus it can be misused as a political slogan.

Before quality of life indicators can be used effectively in the policy arena, we need to refine conceptual categories, test and quantify data, gain the commitment of trained researchers and decision makers, and subject the methods to independent scrutiny. Many items previously thought unquantifiable, such as human intelligence, are now being measured.

Rational environmental decision-making must examine the full range of impacts of actions (including the no-action alternative) on social welfare. The social welfare function may be divided into a number of partly interdependent objectives (economic efficiency, equity, environmental quality, health and safety, and social-political quality of life).

TYPES OF OBJECTIVES

The economic efficiency objective

The economic efficiency objective is a broader measure of economic or material welfare than gross national product or national income. However, GNP is the proxy that is usually used in measures of economic efficiency. Economic efficiency is often described as maximization of "the size of the pie" available to society as a whole.

Strictly speaking, economic efficiency exists when three conditions are fulfilled with regard to efficiency in production, consumption, and exchange. The production possibilities frontier is a hypothetical locus of points representing the maximum amount of goods that can be produced with a given level of inputs (Fig. 8.2).

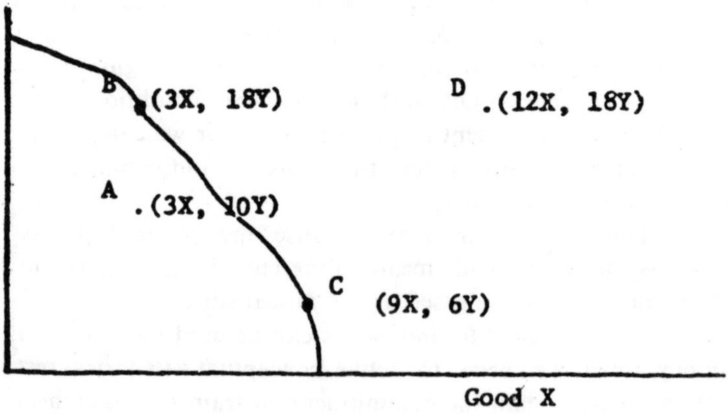

FIG. 8.2 : *The Production Possibilities Frontier*

Points B and C indicate efficient production points. Efficiency requires equality of marginal rates of transformation across all producers. Point A is inefficient since more of good Y could be produced without sacrificing any inputs used in making good X. An inefficiency may result from monopoly or tax distributions. Point D is infeasible since that combination of goods X and Y cannot be produced with the given levels of inputs and technology.

Efficiency in consumption occurs at the tangency between the highest feasible indifference curve and the production possibilities frontier. In other words, the ratios of marginal product to price for all goods must be equal across all consumers. In addition, the marginal rate of substitution between consumer goods must equal producers' marginal rates of transformation for those goods (Fig. 8.3).

Point A is optimal, Point B is inferior and Point C is impossible to achieve.

The third condition is efficiency in exchange. This condition is necessary to ensure that no one's perceived welfare (based on a personal determination by the individual) can be improved by any further trading of goods without making someone else worse off. This state is often termed Pareto efficiency or Pareto optimality. It can be depicted in a simple case by the Edgeworth Box (Fig. 8.4).

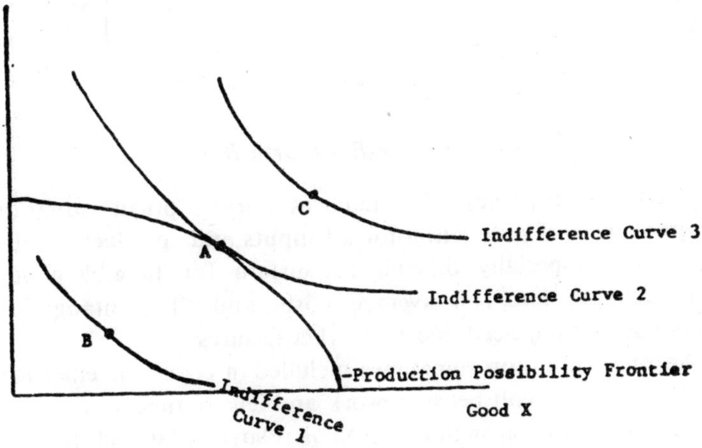

FIG. 8.3 : *Efficiency in Consumption*

The indifference curves concave to the origin O pertain to person M ; those concave to origin P pertain to person N.

The contract curve consists of tangencies between the two consumers M and N. Points off the contract curve are inefficient because voluntary trade will improve one or both consumers. Along the contract curve, all points are Pareto optimal, but movements from one such point to another are not Pareto optimal because they entail someone's gain at someone else's expense. That is why the contract curve is sometimes called a conflict path. The contract curve also implies that economic efficiency is neutral with respect to the distribution of outputs aud resources.

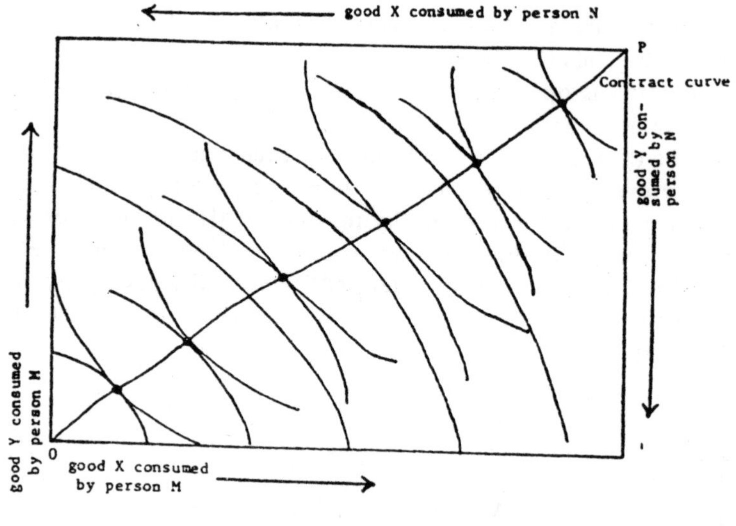

FIG. 8.4 : *The Edgeworth Box*

Economic efficiency also requires correct pricing equal to marginal costs of production for all inputs and products. This condition is especially difficult to sustain for tangible goods with declining long-run average costs and "for intangible" goods that are unpriced due to market failures.

Another condition sometimes included in economic efficiency if an optimal tradeoff between work and leisure time.

The economic growth objective is relatively straightforward

to analyze. Some environmental quality benefits fit into this category, for instance, reduction in damages to marketable outputs or more rational use of productive resources to maximize the present value of net benefits in consumption. The tests of maximum willingness-to-pay for a benefit or minimum willingness-to-accept compensation for a loss apply to economic efficiency impacts. Primary efficiency benefits and costs are real changes in the size of the society's economic pie ; irrespective of their incidence. On the other hand, taxes are transfer payments of pie slices from producers to the government. In a social benefit-cost analysis, taxes should not be deducted from the value of the output. Similarly, subsidies may lower financial costs to individuals, entrepreneurs, or regions, but they do not reduce the social costs of the project. Costs contributed by international agencies should be counted as national economic efficiency costs on the assumption that these subsidies could be used for some alternative purpose within that same country and therefore use of the subsidies implies an opportunity cost.

Shadow prices are necessary in an economic efficiency calculation when market prices do not exist or do not measure social values correctly. The second type of shadow pricing is used most commonly for 1) domestic capital, 2) foreign exchange, 3) labor, and 4) land. In project evaluation, these costs or benefits may be assessed in terms of the product :

(Quantity) (Market Price) $(1 + \lambda)$,

where λ represents the shadow premium.

The shadow price itself is (Market Price) $(1 + \lambda)$. The equivalence of incorporating a shadow price in the objective function or adopting a constraint of the scarce resource can be proven through a simple calculus technique, Lagrangian multipliers.

Capital investment is the engine of economic growth, a society may place a higher or lower weight on the generation of a dollar of capital investment than on an additional dollar of consumption. The extra social value of capital investment depends on the discount rate, the marginal social rate of return on investments, and the relative proportion of returns which are invested rather than consumed. If capital is scarce, it may be given a premium value relative to operating, maintenance, and replacement or labor costs. High interest rates indicate either shortages of local capital and credit, high risk, or inflation premiums.

Foreign exchange commands a premium when it is scarce due to balance of payments deficits or overvaluation of domestic currency in exchange rates. Tariffs may also distort the relationship between domestic prices and world prices. A shadow price assigns more weight to a dollar of foreign exchange cost than a dollar of domestic currency cost. Foreign currency costs incurred after a devaluation of domestic currency are assessed at the new rate.

In general, tradable goods should be valued at international prices for economic efficiency calculations. The relevant price for exports is freight -on-board (f.o.b.) – the world price excluding transportation costs from the border of the exporting country. For imports, the cargo-in-freight price (c.i.f.) should be used—the world price including transportation costs to the border of the importing country.

If unemployment is widespread, government projects can tap labour without reducing total social output. In that case, labor has no opportunity cost and may be considered free because the distribution of wages would be merely a financial transfer payment. In addition to open unemployment where potential workers are visibly unoccupied, several other types of unemployment may exist. Hidden unemployment includes people who have stopped seeking work or have become perpetual students. Disguised unemployment (Nurkse 1954) refers to labor whose marginal produce is zero or negative. With disguised unemployment, marginal workers may be withdrawn without any loss in output. Underemployment consists of people who are employed at a job below their skill level or are only able to find part-time work.

Haveman (1977) used shadow pricing to adjust the economic costs of government projects to account for the probability of drawing unemployed labor as a function of the unemployment rate. His analysis did not distinguish between past and prospective shadow wage rates or differences in unemployment rates between skilled versus unskilled workers and high versus low-wage workers.

Generally, most unemployed or underemployed laborers are performing some kind of work even if it does not show up in markets. A more sophisticated formula defines the shadow wage rate as

$$M+(W-M)-\left(\frac{W-M}{P_K*}\right)$$

where M = marginal product foregone,

$(W-M)$ = increase in consumption by an individual due to extra labor, and

$\left(\frac{W-M}{P_K*}\right)$ = value of current individual consumption in terms of current social savings

(Dasgupta, Sen, and Marglin, 1972)

An additional modification may be made to cover any decrease in the social value of leisure time.

A shadow wage rate should also be used to reflect real social costs if wage payments are above equilibrium levels. For example, governments sometimes pay higher wages than is necessary to hire labor from alternative uses in order to gain political support. Although the extra wage bill does pose a financial cost to the government, it is not an economic resource cost. The equilibrium wage rather than the actual wage would be the shadow price of this labor.

When a public agency already owns land, it may consider the land to be free when it assesses the efficiency benefits and costs of using the site. Nevertheless, this land has an opportunity cost because it could be sold or used in some alternative way. The shadow price of land derives from its highest valued alternative use. Failure to account for the shadow price of land and environmental impacts has led many state highway agencies to conclude erroneously that the cheapest route for construction of highways would pass through state parks.

General inflation affects all segments of the economy equally and it does not change the real relationship between benefits and costs occurring over time. Therefore, the best way to handle general inflation is to estimate all costs and benefits in real dollars in times of some base year and to select an inflation-free discount rate.

In contrast, differential inflation affects certain sectors

disproportionately. The best example of differential inflation is in energy prices which have risen faster than the general rate of inflation. Wheh differential inflation can be predicted, price estimates should be modified by using the difference between the differential and general inflation rates as an escalation factor.

Some price analysts make their economic efficiency estimates more conservative by adding contingency factors. For example, uncertainty might be handled by decreasing expected economic benefits by 20 per cent and/or raising expected costs by 20 per cent. Contingency factors which are proportional to primary costs do not change the relative economic desirability rankings for a group of projects ; however, they might decrease the optimal scale of a project. Contingency factors may be helpful in financial budgeting.

Economic efficiency has limitations as a single objective. Economic efficiency takes tastes and preferences as a given. As a result, economic efficiency has an inherent status-quo bias. Certainly, most people would agree that "merit wants" (goods that individuals choose to consume in quantities less than the socially optimal amounts) and also socially undesirable tastes exist.

Economic efficiency is rooted in the concepts of a free enterprise, perfectly competitive market system. Thus, it fails to deal with market imperfections, common property resources, public goods, externalities, insufficient information, transaction costs, intervention costs, and security costs. Pricing problems are also likely due to indivisibiilties, scale economies and diseconomies, monopoly, and monopony.

Economic indicators used in practice are flaws, still they are well institutionalized in policy making. Economic indicators should emphasize contributions to utility from consumption rather than emphasizing production. Expenditures that do not produce welfare directly could be omitted ; this includes "regrettables" like national defense. External costs such as pollution could be netted out. Leisure and extramarket activities could be added in. GNP would then more closely approximate a "measure of economic welfare" (Nordhaus and Tobin, 1972). As Galbraith noted, if production just satisfies wants it generates, the net welfare product is zero. Others such as Okun argue that GNP should not attempt to measure social welfare and could not tell the whole story under any conceivable re-definition. Welfare cannot be measured exclusively by the quantitative flow of goods and services. We can and should improve economic measures when possible, but this does not obviate the necessity of taking a multiple objectives approach and analyzing how "intangibles" affect social welfare.

The equity objective

The equity objective refers to fairness in the actual distribution of income and wealth and in the processes by which they are distributed. The economic efficiency objective takes the present distribution of goods and services as a given, ignoring their distribution. In a critical value judgement, efficiency-oriented economists assume away a major problem by stating that interpersonal comparisons of utility cannot be made. In fact, it may be desirable to sacrifice some economic efficiency in order to achieve a more equitable distribution. The equity objective can be seen more broadly than just the distribution of income and wealth. It also includes the incidence, access, and actual availability of environmental pollution and of amenities.

Conventional benefit-cost analysis ignores income distribution in its acceptance of the Kaldor-Hicks criterion. That criterion states that social welfare is improved when total benefits exceed total costs regardless of their incidence. Thus, it would be possible for gainers to compensate losers for their losses and still be better off than before. This compensation does not have to be paid ; it just must be possible. In economics terms, this is called a potential-Pareto change. The Kaldor-Hicks criterion implies that the total size of the economic pie is the deciding factor while the relative shares can be apportioned in any way. That criterion originates from the utilitarian concept of social welfare as the unweighted sum of individual utilities.

Clearly, most definitions of development (as opposed to economic growth) recognize the distribution of income and wealth as a determinant of social welfare. Equity becomes crucial in countries which have a large amount of relative poverty (unequal apportionment of economic resources) as well as absolute poverty (material deprivation). Public investments can be an effective way of redistributing income without placing stigma on the recipients whereas simple transfers of income will incur more political opposition. A multiple-objectives approach explicitly incorporates distributional effects in the objective function for public project evaluation. In measuring equity, we really are interested in wealth (the stock equivalent of income). However, the poor rarely have much stored wealth and wealth would be very difficult to measure. Income is used as a surrogate for wealth because it is easier to estimate.

Relative poverty is defined by adopting some statistical poverty line, often an arbitrary percentage of the population such as the lowest-earning 20 per cent or 40 per cent. Absolute poverty is harder to measure than relative poverty. First, the analyst must estimate a minimum level of consumption based on either biologically or culturally-determined needs. Second, the analyst computes the amount of income necessary to support that level of consumption, accounting for subsistence production by the household. According to Sen (1978), a measure of poverty should reflect both the "poverty gap" (absolute shortfall between household income and the poverty line) and the "poverty subset" (number of people experiencing poverty). Ideally, the index should also improve when income is transferred from one person below the poverty line to another who is even poorer. The marginal social utility of income can be defined as a function of the income of the recipient. Generally, some reference income level Y* is established. Assuming a constant marginal social utility of income ε the distributional weight or shadow price of income d_i for income group y_i will be

$$d_i = \left(\frac{y^*}{y_i} \right)^{\varepsilon}$$

(Lal 1974, p. 43).

Freeman (1977, p. 253) provides seven hypothetical weighting functions including the parameter α or $-\varepsilon$, given a base annual income of \$10,000 and value weights placed on annual incomes of \$1,000 ; \$100,000 ; and \$1,000,000. For example, when $\alpha = 1.0$ or $\varepsilon = -1$, the marginal utility of income at half the poverty line would be two and it would be one half when income is twice the poverty level. This distributional weight is equivalent to the one recommended by Squire and Van der Tak, "For most governments, *n* [α], would probably center around one. Values close to zero or two, although possible would be extreme" (1975, p. 63). In order to avoid double counting of efficiency and equity benefits, the shadow premium of income (λ_y) equals $d_i - 1$. Rather than computing separate value weights for each possible level of income, income classes can be defined and represented by the mean income within the class.

The regional development objective

It is not clear whether regional development is a separate objective from equity in the distribution of income and wealth to individuals and groups. Care must be taken to avoid double counting these benefits as national income benefits and equity benefits. Equity effects are correctly reflected and there is no double counting problem if (1) a depressed region has unemployed resources that cannot be moved to other areas, (2) the project affects regional income directly through output or indirectly through changes in economic activity, (3) one accounts for displacement effect losses that occur elsewhere as a result of withdrawing resources from other locations, and (4) one does not make the tenuous assumption that benefits in a depressed region trickle down from project beneficiaries to low-income residents.

When primary benefits can be measured in dollar terms, the question of whether to count secondary benefits arises. This is also the case for environmental improvements that increase the productivity of market outputs from farmland, fisheries, forests, and other natural resources. The issues of secondary benefits is related more to regional development than to economic efficiency.

Primary (direct) benefits equal the value of the extra (marginal) production associated with a project. Secondary benefits are increases in the income of factors of production beyond the primary output of a project. Often, secondary benefits show up as profits for related activities. Secondary benefits are often indicated by phrases like "value added" or "activities stemming from or induced by a project". For example, salmon population in a river may rebound after improvements in water quality. The increased value of fish at dockside is the primary benefit ; higher profits for salmon canneries or worm suppliers are secondary benefits. Similarly, greater net revenues for wheat farmers on irrigated land is a primary benefit while the activities of local wholesalers and retailers handling farm goods from irrigated land, industries supplying farm inputs or farm household items, and other enterprises between the farm and final consumer represent secondary benefits.

In many instances, huge secondary benefits have been tabulated in order to justify wasteful projects. In general, secondary benefits should not be counted as economic efficiency benefits to the nation or a region. In analyzing benefits to a local area, it may be legitimate to account for secondary benefits if opportunity costs are netted out.

One of the reasons why secondary benefits are often spurious rests on the distinction between technological and pecuniary externalities. A technological externality is a real economic efficiency loss or gain caused by spillover effects from the actions of some producers or consumers on others. Technological externalities affect the physical productivity of resources or the enjoyment of consumption, figuratively speaking. They change the size of the economic pie. In contrast, pecuniary externalities are merely transfer payments from one group to another ; they just redistribute the size of pie slices to various groups while the pie size remains constant. In fact, many secondary benefits are transfer payments from one industry, firm, or household to others.

Most of the time when secondary benefits are counted, secondary costs are forgotten. Expanding output in some sector usually means withdrawing resources from other uses and these shifts present opportunity costs. Frequently, on a regional or national scale, secondary costs cancel secondary benefits. Looking through a magnifying glass at all of these effects becomes impossible, so it is probably not worthwhile to trace secondary benefits and secondary costs.

The only time it is justifiable to consider secondary benefits is if there are immobile unemployed resources without an opportunity cost or if there are economies of scale or aglomeration. However, most resources are mobile, in general. New places may grow at the expense of others. Although projects can have positive effects on less developed areas, it is better to separate the efficiency benefits from the redistribution benefits. Then these effects can be counted as regional development benefits.

Income multipliers are often used as a simple, but incorrect, way of determining secondary benefits. For instance, if a project produces $100 worth of income through primary benefits, it can stimulate the economy in later time periods by increasing aggregate demand. If the marginal propensity to consume is 0.5, the $100 will generate $50 worth of spending in period 2, $25 in period 3, $12.50 in period 4, etc. Eventually, the secondary effects approach zero. The effect in the nth year is $100 \times 0.5^{n-1}$. The sum of the additions to income in all of the subsequent rounds beyond the initial injection of money adds up to $\frac{K}{1-C}$, where K is the initial amount of primary benefits, and C is the marginal propensity to consume. In the example above, multiplier respending generates $100 in addition to the initial $100. The multiplier method is faulty because it fails to consider what would have happened to incomes if the project had not taken place. The financial or physical resources consumed by a project could have been used in alternate ways to generate multiplier income effects. In evaluating benefits, we are only interested in differences between the "with project" case and the "without project" case. It can be shown that inclusion of multiplier effects does not alter the ranking of project worth.

	Alternative A	*Alternative B*
Primary Benefits	15	12
Multiplier Effects (C=0.5)	15	12
Total Benefits	30	24

Considering the primary benefits only, alternative A is preferable. Adding the multiplier effects does not change the relative desirability of alternatives because the multiplier is proportional to the primary benefits. Inclusion of multiplier effects will boost the total present value of net benefits of each project including the no-action alternative. When the present value of net benefits is greater than one only when multiplier effects are included, a project is submarginal and the opportunity cost of consuming the resources to carry out the project probably exceeds the benefits.

The environmental quality objective

Ecological factors should be viewed as a basic objective in decision making rather than a mere constraint to economic efficiency. Environmental quality is recognized as an important societal objective by the general public, experts, and the law. The sharp increase in environmental pollution between 1950 and 1970 spurred a public response that led the U.S. Congress to adopt major legislation (NEPA 1969, CAA 1970, FWPCA 1972, and Endangered Species Act of 1973, Toxic Substances Control Act of 1976, and the Surface Mining Control and Reclamation Act of 1977, among others.

Environmental quality has to be listed as a separate objective because economic theories of externalities and collective goods provide only a limited operational framework for incorporating all the dimensions of environmental quality in an economic efficiency calculus. To the extent that environmental quality can be evaluated in economic terms, one should do so. However, it is impossible to incorporate all dimensions of this objective in the economic model. There is no single satisfactory numeraire of "environmental quality units" analogous to money prices for market goods and services. (Hufschmidt and Hyman, 1981 ; Moreau, Hyman, Stiftel and Nichols, 1981.)

The health and safety objective

Public health and safety has long been a basic concern of society. Numerous laws dealing with environmental quality, as defined broadly, were designed to protect health and safety of individuals of all ages and group affiliations, Environmental impacts on health and safety are very important.

The social quality of life objective

Most importantly, economic efficiency is only one component

of a social welfare function. Social and political factors and processes are very important as well. Philosophically, more material goods may not always be better for society. Preferences can be ascertained for a variety of social aspects that cannot be quantified in strictly economic variables, including kinship, friendship, and social status ; health ; time and leisure ; employment and quality of working life ; individual development through learning ; personal safety and the administration of justice (Organisation for Economic Cooperation and Development 1976). Social indicators can measure attainment of quality of life objectives.

Many of these indicators are of a non-economic nature. However, some of these goods and services do have market components and others potentially are marketable, but remain extramarket because they are unpriced in the existing institutional structure. Frequently, these goods and services are classified as merit wants or equity wants. Merit wants are consumed by individuals in lower quantities than society deems optimal. Similarly, society presumes that everyone is entitled to a minimum quantity of equity wants to satisfy basic needs and principles of fairness. Community participation and institution building are ways of ensuring provision of these goods.

FORMULATING AND WEIGHTING OBJECTIVES

Societal objectives are formed in many ways through social norms, professional norms, and political participation. In practice, they are formulated by planners, administrators, and decision makers in executive agencies and bureaus, legislators, judges, and even low-level implementers.

After the objectives have been determined, the next step is to define specific indicators for the evaluation of each objective. Then, relative importance coefficients (RIC) must be assigned to each indicator within an objective. In other words, the relative social welfare weight attached to an objective is partitioned to the subobjectives.

Now, in order to add up effects within an objective, the indicators must fit on the same measurement scale. When all of the indicators within an objective are measured in dollars, a common numeraire already exists. However, some of the indicators can be expressed in qualitative terms only. The solution is to convert the various measuring rods into zero-one scales. This conversion is legitimate because only the relative values affect rankings.

One way of scaling is to convert the raw scores into percentages of the best score in the sample and then divide by one hundred. For example, if five projects have present values of net economic efficiency benefits of $1000, $2500, $3000, $4500, and $5000 respectively, the corresponding scaled scores would be 0.20, 0.50, 0.60, 0.90, and 1.00. This first method can be used when the highest possible logical score cannot be defined. Conceivably, the monetary returns to a project could be infinite.

If any of the raw scores happens to be negative, it becomes necessary to compute the highest absolute value of the sample scores and use that as the base figure. When this complication occurs, the scale ranges from −1.00 to +1.00. A negative score may be interpreted as an economic cost. As an example, the scaled values of −1000, −50,0, +200, +360, in order, become −1.00, −0.05, 0.00, +0.02, +0.36.

Similarly, each raw score can be converted into a percentage of the best possible score, when that score can be defined. This second method is most applicable for qualitative indicators. Give the following hypothetical scores for 0 to 3 : nil (0), low (1), low (1), medium (2), and high (3), the corresponding scaled values would be 0.00, 0.33, 0.33, 0.50, and 1.00.

To summarize, define $INDIC_i$ as the scaled indicator score. Assuming that there are *n* indicators within an objective, the total score, O_J, for the objective can be represented as

$$O_J = \sum_{i=1}^{n} (INDIC_i)\,(RIC_i).$$

Given scaled scores and social welfare weights for each of the *m* objectives, one can compute the social welfare index, *U* :

$$U = \sum_{j=1}^{m} (O_J)\,(WT_J)$$

Once weights are obtained, there are several equivalent forms that an objective function may take. In the first form, the weights are implicit in constrained maximization. The first form involves maximizing net economic efficiency benefits subject to constraints of other objectives.

Maximize $[Y-M-K]$

subject to 1. $Y = f(X_1, X_2, \ldots, X_n)$
2. $D > D^*$
3. $E > E^*$
4. $H > H^*$
5. $S > S^*$

where Y = present value of gross economic efficiency benefits
M = present value of operating, maintenance, and replacement costs
K = present value of capital costs
X_i = inputs
D = present value of distributional benefits
E = present value of environmental quality benefits
H = present value of health and safety benefits
S = present value of social quality of life benefits
$*$ = desired minimum level.

Alternatively one could maximize nonefficiency objectives subject to an economic efficiency constraint:

Maximize $[W_2D + W_3E + W_4H + W_5S]$
subject to 1. $Y = f(X_1, X_2, \ldots, X_n)$
2. $(Y-M-K) > (Y-M-K)^*$

where W_i = relative weights on the importance of objectives over that range of goals-achievement.

The third form involves maximization of the entire objective function:

Maximize $[W_1Y + W_2D + W_3E + W_4H + W_5S]$
subject to 1. $Y = f(X_1\ X_2, \ldots, X_n)$

Ideally, a social welfare function with accurate weights for different objectives over different ranges of goals achievemnt could be determined. However, this is not likely to be the case. The best that we can hope for is some consensus on basic decisions and an explicit presentation of weights so that they can be challenged by opposing groups. Weights are generated implicitly in the day-to-day processes of societal decision making, but these weights are not necessarily rational or consistent.

Multiple-objectives planning in practice

One of the earliest conceptual treatments of the multiple-objective approach was in the field of water resources. The Harvard Water Program (Maass, Hufschmidt, et al. 1962) developed a general design methodology capable of including evaluation criteria other than those specifically measured in economic efficiency terms. This work strongly influenced the issuance of Senate Document 97 (1962) which established new standards for the formulation and review of water and related land resource development. This policy document listed three planning objectives : Development, Preservation, and Well-being of People.

The "problem" of planning for multiple objectives was also addressed in the field of urban and regional planning. Lichfield (1960), unsatisfied with the limitations of benefit-cost analysis, extended the method to include noneconomic categories in a "planning balance sheet". Hill (1973) developed a more sophisticated "goals-achievement matrix" which integrated monetary and non-monetary aspects in the evaluation of alternatives. This approach estimates the benefits and costs of each alternative, not in monetary terms, but in terms of their relative contribution to the total array of objectives. The method uses a weighting scheme to reflect a community's importance rating of various objectives as well as its preferences for the incidence of benefits and costs on various individuals and groups.

In September of 1973, the U.S. Water Resources Council adopted a variant of the multiple-objective approach for planning water and related land resources. These *Principles and Standards* (1973) recognize the multiple objectives nature of public policy by establishing four accounts : National Economic Development (NED), Environmental Quality (EQ), Regional Development (RD), and Social Well-Being (SWB).

However, *Principles and Standards* allows for calculations to be made on the NED and EQ accounts as full objectives. RD and SWB are treated as display objectives ; although they are outside the formal, analytical decision-making calculus, decision makers are allowed to consider effects on these objectives in reaching a decision. Project formulation may be based only on NED and EQ objectives. When effects on the Regional Development and Social Well-Being accounts are only "displays" and not full objectives, they tend to receive insufficient consideration both in the design and evaluation stages.

Principles and Standards is not a fully satisfactory treatment of the objectives, as currently defined. The accounts are not mutually exclusive, and there is some confusion about the proper categorization. The NED account is internally consistent as it is based on national economic efficiency. However, the RD account reflects national economic efficiency as well as regional equity, hence this can lead to double counting.

The EQ account appears to be limited to natural system effects and values. Environmental quality effects are measured in all of the *Principles and Standards* accounts.

The SWB account is a catchall. It includes health and safety factors and amenity values even when they are related to environmental quality. Little has been done to extend the SWB account into the full range of other aspects of the quality of life. Social indicators can be used to gain additional information to measure social well-being.

Table 8.1 compares the *Principles and Standards* accounts to the more general objectives discussed above. This distinction is particularly important with regard to the environmental quality objective which contains diverse components dealing with natural ecosystems ; economic productivity of resources ; human health and safety ; amenities and aesthetics ; and historical, archeological and cultural aspects. A matrix that displays effects across a series of objectives can be very useful in the evaluation of alternative projects and plans.

Nevertheless, *Principles and Standards* is a very important document because its regulations apply to project formulation and evaluation by U.S. federal agencies concerned with water and related land resources. In addition, its scope has influenced environmental impact assessment and multiple-objective plan-

TABLE 8.1 : Planning objectives set forth in WRC principles and standards accounts

Types of National Objectives	Basic Accounts in Principles and Standards		Supplementary Accounts in Principles and Standards	
	National Economic Development	Environmental Quality	Regional Development	Social Well-Being
National Economic Efficiency	Almost Entirely	—	Partly	—
Regional Equity	Partly	—	Partly	—
Individual and Group Equity	—	—	Indirectly	—
Environmental Quality	Partly (recreation, commercial, fisheries, etc.)	Partly (amenities and aesthetics, natural systems, historic, archeological and cultural values)	Partly (incidence and distribution of environmental quality)	Partly (environmental health and morbidity and mortality)
Health and Safety	Partly (economic losses due to reduced human capital)	—	Partly (incidence)	Partly (morbidity and mortality—personal and external effects)
Other Aspects of Quality of Life	—	—	Partly (incidence)	Partly

ning in other sectors. Its significance is demonstrated in many of the methods of environmental assessment (Nichols and Hyman, 1982 ; Hyman, 1981) and in the practice as well.

REFERENCES

Banfield, Edward. "Note on a Conceptual Scheme", in *Politics, Planning and the Public Interest*, by Martin Meyerson and Edward Banfield, New York : The Free Press, 1955.

Campbell, Angus, P. Converse, and W. Rodgers. *The Quality of American Life*, New York : Russell Sage Foundation, 1976.

Dasgupta, Partha, A.K. Sen, and Stephen Marglin. *Guidelines for Project Evaluation*, New York : United Nations, 1972.

Freeman, A. Myrick III. "Project Design and Evaluation with Multiple Objectives", In *Public Expenditure and Policy Analysis*, 2nd ed., edited by Robert Haveman and Julius Margolis, Chicago : Rand McNally, 1977.

Haveman, Robert. "Evaluation of Public Expenditures under conditions of Unemployment", in *Public Expenditures and Policy Analysis*, 2nd ed., edited by Robert Haveman and Julius Margolis, Chicago, 11 ; Rand McNally, 1977.

Hill, Morris. "A Goals-Achievement Matrix for Evaluating Alternative Plans", *Journal of the American Institute of Planners* 34 (1968) : 19-28.

Hufschmidt, Maynard. "Environmental Quality as a Policy and Planning Objective", *Journal of the American Institute of Planners* 4 (1971) : 231-242.

Hufschmidt, Maynard. *Environmental Statements and Water Resources Planning in North Carolina*, Raleigh, NC: Water Resources Research Institute, University of North Carolina, 1974 (NTIS PB-235, 265).

Hufschmidt, Maynard and Eric Hyman, eds., *Economic Valuation of Natural Resource and Environmental Aspects of Development.* Wicklow, Ireland : Tycooly Press (UNIPUB), 1981.

Hyman, Eric. "The Valuation of Extramarket Benefits and Costs in Environmental Assessment", *Environmental Impact Assessment Review*, 1981.

Janis, Irving and Leon Mann. *Decision-Making: A Psychological Analysis of Conflict, Choice and Commitments*, New York : The Free Press, 1977.

Lal, Deepak. *Methods of Project Analysis : A Review*, Baltimore, MD : Johns Hopkins University Press, 1974.

Lichfield, Nathaniel. "Cost-Benefit Analysis in City Planning", *Journal of the American Institute of Planners* 26 (1960) : 273-279.

Maass, Arthur, M. Hufschmidt, R. Dorfman et al. *Design of Water Resource Systems*. Cambridge, MA : Harvard University Press, 1962.

Mishan, E.J. *Technology and Growth : The Price We Pay*, New York :

Praegar, 1969.

Moreau David Eric, Hyman, Bruce Stiftel, and Robert Nichols. *Elicitation of Environmental Values in Multiple-Objective Water Resource Decision-Making*, Springfield, VA : National Technical Information Service, 1981.

Nichols, Robert and Eric Hyman. "An Evaluation of Environmental Assessment Methods", *Journal of the American Society of Civil Engineers, Water Resources Planning and Management Division*, 1981.

Nordhaus, P. and James Tobin. "GNP as a Measure of Economic Welfare", *Economics of the Environment*, edited by Robert and Nancy Dorfman, New York]: W.W. Norton, 1972.

Nurkse, Ragnar. "Bxcess Population and Capital Construction", in *Leading Issues in Development Economics*. edited by Gerald Meier, New York : Oxford University Press, 1964.

Organisation for Economic Cooperation and Development. *Measuring Social Well-Being*, Paris : OECD, 1976.

Quade, Edward. *Systems Analysis Techniques for Planning, Programming, and Budgeting*, Santa Monica, CA : Rand Corporation, 1966.

Sen, Amartya. *Notes on the Concept of Poverty*, Geneva : International Labour Organization, 1978 (World Environment Program Working Paper, No. 65).

Squire, L. and H. Van der Tak. *Economic Analysis of Projects*, Baltimore, MD : Johns Hopkins Press, 1975.

Stiftel, Bruce and Eric Hyman. "Assessment of Environmental Quality in a Democratic State", *The Environmental Professional*, 2 (1981) : 306-314.

United States Senate. *Policies, Standards, and Procedures in the Formulation, Evaluation, and Review of Planning for Use and Development of Water and Related Land Resources*. 27th Congress, Senate Document No. 97, 1962.

U.S. Water Resources Council. "Water and Related Land Resources : Establishment of Principles and Standards for Planning", *Federal Register* 38 (September 10, 1973) : 24778-24869.

U.S. Water Resources Council. "Final Rule : Principles and Standards for Water and Related Land Resources Planning", *Federal Register* 45 (September 29, 1980) : 64366-64400.

NINE

R.S. DORNEY AND E.S. HIGGS

URBAN VALLEYLAND CLASSIFICATION FOR OPEN SPACE PLANNING

INTRODUCTION

THE late 1950s and early 1960s witnessed a series of publications that highlighted aspects of deteriorating regional and urban environmental quality in a fashion that captured the public's attention (McIntosh, 1976). These publications formed the initial impetus for a rise of environmental consciousness. The general support for better environmental quality was linked to public health and safety. This so-called ecological thinking referred not only to a greater concern for the natural environment but also to the notion that interactions between nature, man, and his technology were dynamically interrelated.

The study of ecology as a natural science blossomed at this time both in North America and Europe (Curry-Lindahl, 1978). Accompanying the rise of ecological thinking in the 1960s and 1970s was a growing public acceptance of anti-growth, anti-technocracy and anti-urbanization concepts (Dorney, 1977a). Although this new concern left many politicians, scientists and planners unprepared, agencies and some professions were quick to react to the situation; the United Nations Environment Programme, United States Environmental Protection Agency, the Federal Canadian Department of the Environment and the Canadian Institute of Planning, to name but a few. Organizations in government and private sectors evolved capabilities necessary to undertake projects with environmental components.

This period spawned new companies specializing in ecological and environmental consulting, while many academic centres developed environmental studies programs.

As one would expect, this movement produced a response, initially in the universities and in the private sector to developing landscape classification methodologies. Botanists such as Numata developed urban biotic maps based on plant community structure (Tokyo, 1974) while geologists such as Christiansen (1970) mapped substructures of cities. Maps based on energetics were developed (Dansereau and Pare, 1977). In addition, synthetic maps of various kinds showing value surfaces (e.g. high value, medium value, low value) appeared (e.g. McHarg, 1969 ; Dorney and Priddle, 1973) whose principal objective was to improve the man-nature fit through design. Next, these synthetic maps were followed by development of computer graphics with complex software programs designed to solve planning problems (e.g. Fabos, 1981).

Somehow in this burgeoning field of landscape mapping, the user was ignored. It was assumed that excellent field science, incorporated into increasingly more complex graphics would lead to better and better land use decisions. It is now painfully apparent to some practitioners at least that more is not better, and that it is paramount to first define the user. Then it is necessary to tailor the graphic system to the user rather than expect the user to spend time learning a particular graphic system. It is this user orientation which was explicitly pursued in developing the methodology presented here.

In this paper we will be presenting a three-map descriptive system or methodology for classifying urban valleyland areas. Valleylands are particularly sensitive zones to urban intrusion and alteration. They can be cold air drainage channels, they can be hazardous areas because of flooding and slope instability, they may serve to carry storm and sanitary water as well as freeways or arterial roads. In addition, they have passive and active recreational potential and serve to link urban elements. Because of these multiple uses, conflicts between existing and potential land use priorities arise frequently.

Conservation Authorities, as they are known in Ontario, or other river valley authorities responsible for monitoring and managing watersheds, may impose physical boundaries that attempt to limit building development on floodplains or in adjacent upland areas. A popular expression of this boundary approach has been the establishment of so-called "flood and fill lines". These lines delimit where hypothetical storm flooding peaks might occur and delineate as well the top of banks beyond which fill is not permitted. The fill line used here, then becomes a convenient outer boundary for valleyland mapping and for urban planning.

A recent Ontario approach to the planning of regional and urban valleylands as open space has been the Environmentally Sensitive Policy Area system (Dorney and George, 1970 ; Francis and Eagles, 1975 ; and Adindu and Eagles, 1977). The impetus for this approach, which stresses the importance of small-scale regionally significant natural features, began in the Regional Municipality of Waterloo in 1970 and includes protection of agricultural land, and mineral resources (Dorney and Hoffman, 1979).

Aside from policy planning and associated zoning approaches, economic justification can often be developed for the protection of urban valleylands. This comes about in three ways :

1. Property values can be enhanced if natural features such as valleylands, trees, are retained.
2. Knowledge of certain natural processes may provide direct financial savings to both the developer and the home owner in terms of maintaining a water table, in preventing erosión, or in protecting unstable slopes, etc.
3. Information developed in urban design on biophysical and cultural features can prevent delays in development approvals.

The general methodology to be presented in this paper was developed for Townsend, a "new town" planned for a rural area in Southern Ontario (Dorney, 1977b). It utilized a 3-map approach of abiotic, biotic, and cultural features (abc mapping system) which conceptually develops what is an opportunity-constraint approach for urban development by depicting existing rural areas which can be developed easily and those which should not or cannot. Later for the City of Waterloo an urban biotic classification system was published utilizing physiognomic characteristics (Brady *et al.*, 1979). Then these two mapping systems were combined so that the urban valleylands in both rural areas and in built-up urban areas were mapped (Dorney and Higgs, 1978). This mapping system can be used to map the entire surface of cities as well as just the valleylands.

The methodology

In designing the study methodology four primary concerns were addressed : (*i*) the data collected must have strength before judicial bodies, (*ii*) the data collection should be restricted as little as possible by season, (*iii*) the methodology should be useable by study groups with small budgets, and, (*iv*) the methods employed must be thorough and well documented to allow other observers to reproduce the data. The introduction of these concerns into the methodology was accomplished by taking the following into consideration :

(*a*) a systems approach,
(*b*) an iterative rather than linear approach,
(*c*) a user orientation, and
(*d*) a series of technical principles.

(*a*) The systems approach provides an integrative framework for the organization and manipulation of the methodology. It allows the superimposition or overlapping of abiotic, biotic and cultural components or subsystems to give a coherent picture of the functional aspects of any given area. Our methodology considers the valleylands as open systems but not tending toward unlimited entropy or randomness. In ecological systems, a negative entropy or greater organization can be achieved through energy exchanges between the system (valleylands) and its external environment. The three subsystems are convenient study divisions which assist the study team in data collection and synthesis.

(*b*) An iterative methodology allows considerably grerate flexibility to both the study team and the users in fitting the methods to a particular project. No methodology is able to meet the requirements of all projects in a template-like manner ; there must be room for modifications in the methodology as the study progresses. For example, if it were found that the presentation maps were to be inadequate for clear interpretation by a large number of users, the style or content of the maps could be changed while the project is ongoing.

(*c*) The potential user groups in our methodology are identified, and the primary issues of concern to them outlined. Potential users include engineers, planners, recreationists, residents, naturalists, historians, farmers, fishermen and scientists. A list of user concerns, in the form of questions were developed to assist a study team in meeting the needs of specific users. These user concerns, which may vary from one region to another, have been omitted for reasons of space but are found in the full report (Dorney and Higgs, 1978).

(*d*) The 42 technical principles developed earlier (by Dorney and Hoffman, 1979) encompass a range of concepts to be considered in any environmental undertaking, such as productivity, indicator species or low risk design. They assist in developing a study design, timetable and budget.

The field studies, and mapping were designed to be undertaken by a minimum of three experts ; one each for the abiotic, biotic and cultural/aesthetic subsystems.

Data collection

The collection of baseline data is designed as discussed above to develop an easily understood three map series : abiotic features, biotic features, and cultural/aesthetic features (Figures 9.1-9.4). This three map series (called the abc method) generally needs to be done for urban areas at a scale of 1″ to 400′ or 1″ to 200′ (1 : 5,000, or 1 : 2,500). The level of detail at 1 : 5,000 scale seems to be appropriate for most urban ecosystem studies. The 100 year flood lines and/or the top of valley walls plus a 100 feet or 30 meter setback can form the boundary of the valleyland mapping system.

The abiotic subsystem includes (*i*) slopes over 15%,

(*ii*) surficial geology, (*iii*) soil types, (*iv*) air quality data and (*v*) water quality data.

Slopes over 15% are particularly significant as constraints to development. Slopes of this (class) also have recreation potential for sledding, and have aesthetic appeal, forming enclosed valleys. The designation of these slopes can be made from existing agricultural soils maps in the field or from city contour maps.

The surficial geology maps are used to describe the major surficial land forms : kames, till plains, lacustrian deposits, and man-made fill. These are mapped directly and listed in the legend.

Soils mapping is transferred directly from regional maps to city scale maps. Where engineering drainage was known to have occurred in the city, the drainage class of the soils was changed. In this event, each soil type should be given an appropriate abbreviation. Soil types are overlain on the surficial geology maps (Fig. 9.2).

Water and air quality sampling and analysis can be done in the field, or if available, it can be compiled from existing data sources. Many cities now have such inventory information. None is shown on these maps.

Biotic resources can be classified from both natural and cultural vantage points (Dorney and Hoffman, 1979). The vantage point used here is to describe the vegetation units based on physiognomic structure (Brady *et al.*, 1979). Using this approach the urban surface is divided into 15 landscape units based on the 18 units published by Brady *et al.*, 1979. This typology describes the biotic surface to one-quarter acre minimum size (Figu. 9.3). Vegetation mapping requires current aerial photos or photo mosaics, with field checking. Wildlife observations can accompany this vegetation mapping with or without resorting to quantitative breeding bird or wintering bird transects or trapping of mammals.

Cultural/aesthetic data collection is set in three parts : the cultural/historic features, aesthetic features and existing and potential recreation sites (Fig. 9.4). Cultural/historic data are collected through readings, interviews and field observations. Buildings of historic interest can be plotted on maps. These features assist in determining recreation potential and the degree of effort required for rehabilitation of a particular reach of the stream. Criteria that assist in the aesthetic analysis of valleylands include viewpoints, historic buildings and sites viewpoints, degree of valley enclosure, adjacent amenity/ disamenity land uses, recreation sites, slope condition, presence of wildlife, wildlife concentrations of particular interest to the viewing public, water and air quality, disamenities and unsatisfactory noise levels. We have found the aesthetic evaluation procedures of Leopold (1967), Sargent (1965), Lewis (1969), Linton (1968), and Maniate and Carter (1973) more applicable to regional than to urban landscape problems, hence have developed the factors outlined above.

Synthesis maps

For the most part, the preparation of synthesis maps is a potentially useful but expensive step. If the focus is on recreation suitability a different set of criteria will be needed than for producing a map intended to reduce park maintenance costs. For this reason we see the three map abc series as useful baseline information for a variety of users who will prepare their own synthesis maps using criteria specific to their need. If the more expensive computer graphics approach is desired, the abc maps could form the basic data needed for providing information digitized prior to scenario development or modelling.

Results

Figures 9.2, 9.3 and 9.4 depict a sector of the City of Waterloo showing the three map series with the legend shown as Figure 9.1. For this City of 50,000 people, a total of 15 sectors (45 maps) were required at a scale of 1″ to 400′ (approx. 1 : 5,000). The fragmentation of the biotic features typical of urban areas is quite evident.

The systems aspects of the mapping, or the ecological process and dynamics must be inferred from the knowledge of scientists, designers and engineers familiar with the regional biophysical landscape. The least expensive way to do this is by using the ad hoc interdisciplinary team approach (Dorney, 1973) relying either on a local citizens advisory team, an in-house agency team, or consultants.

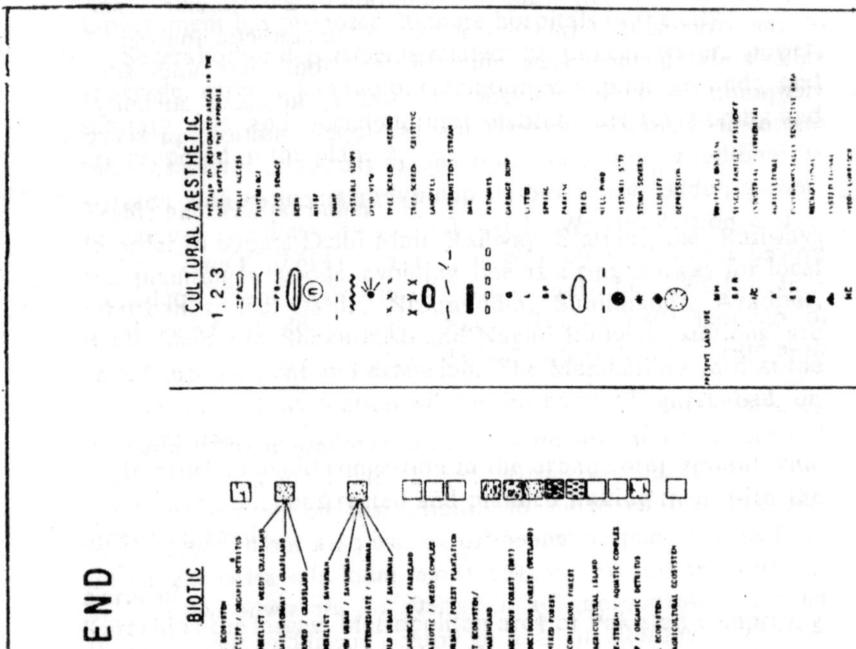

FIG. 9.1 : *Legend for Figures 9.2, 9.3, 9.4*

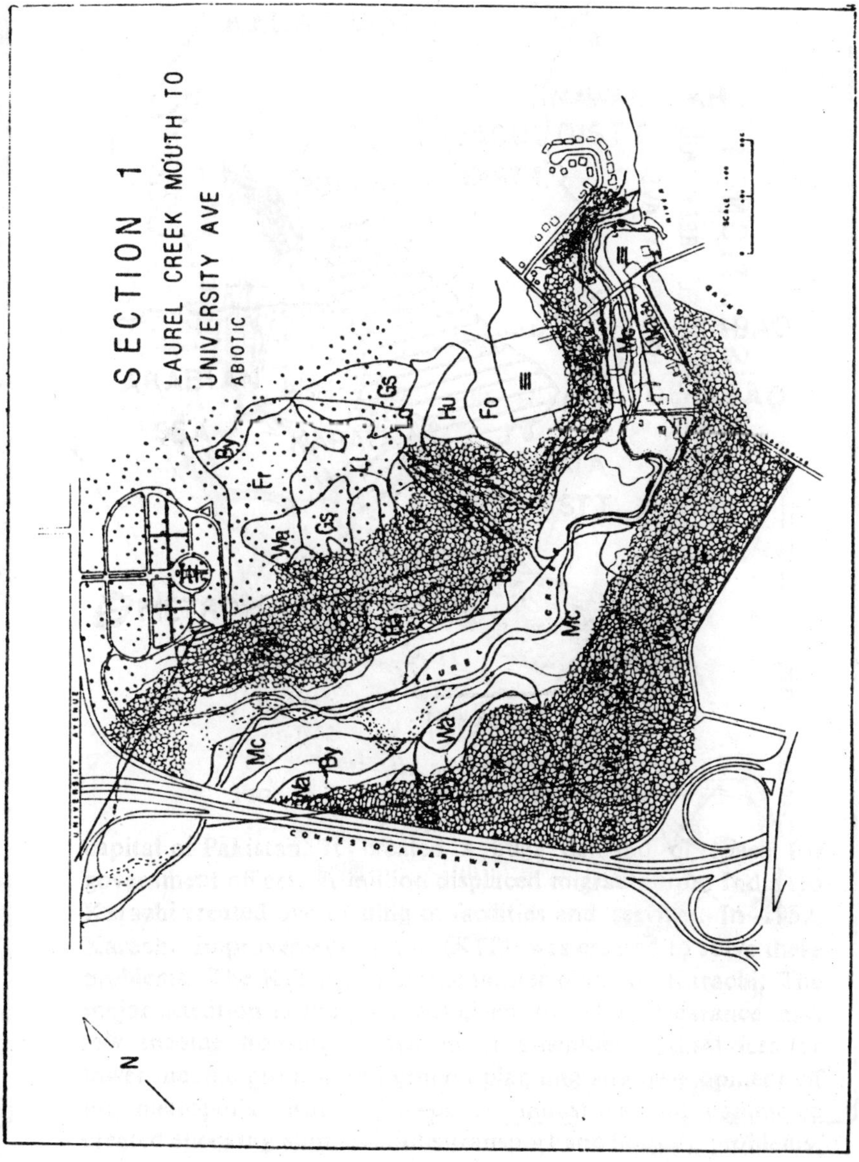

FIG. 9.2 : *Representative Section of Abiotic Surface of the City of Waterloo*

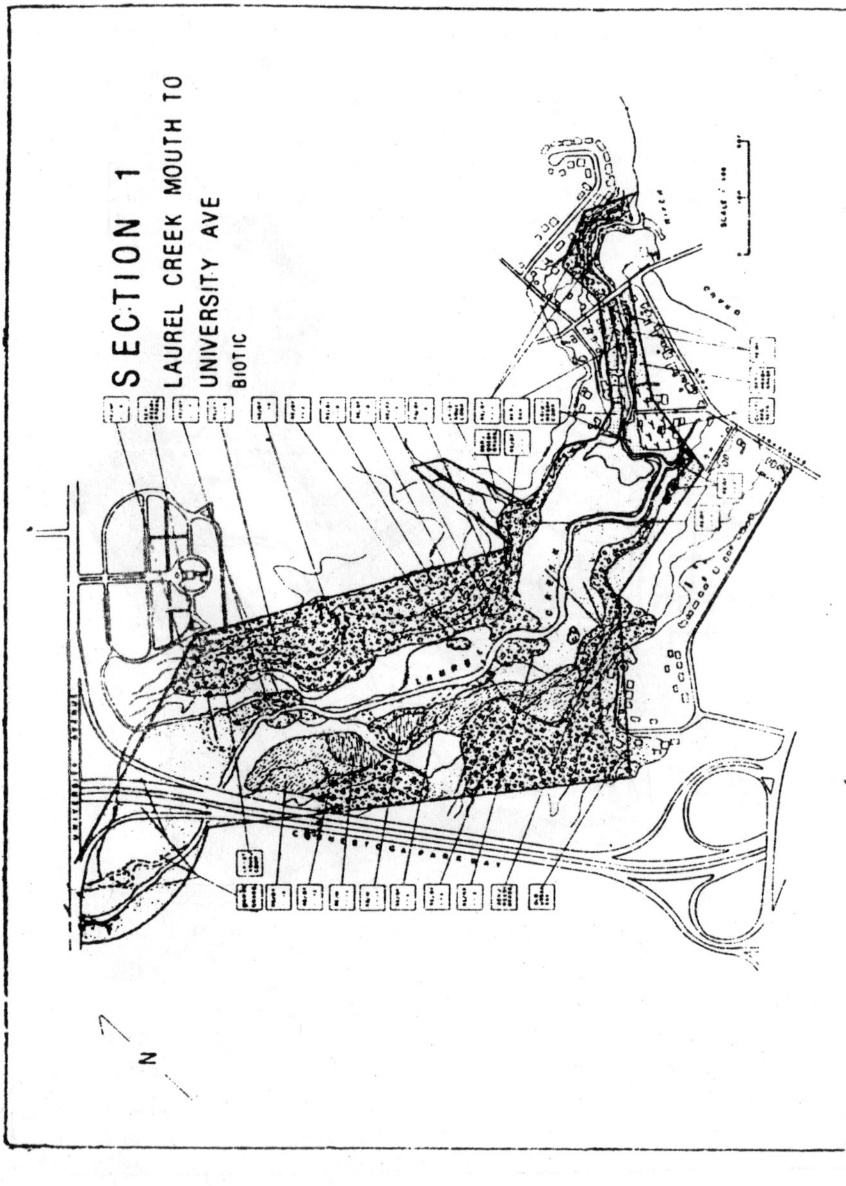
SECTION 1
LAUREL CREEK MOUTH TO
UNIVERSITY AVE
BIOTIC
N

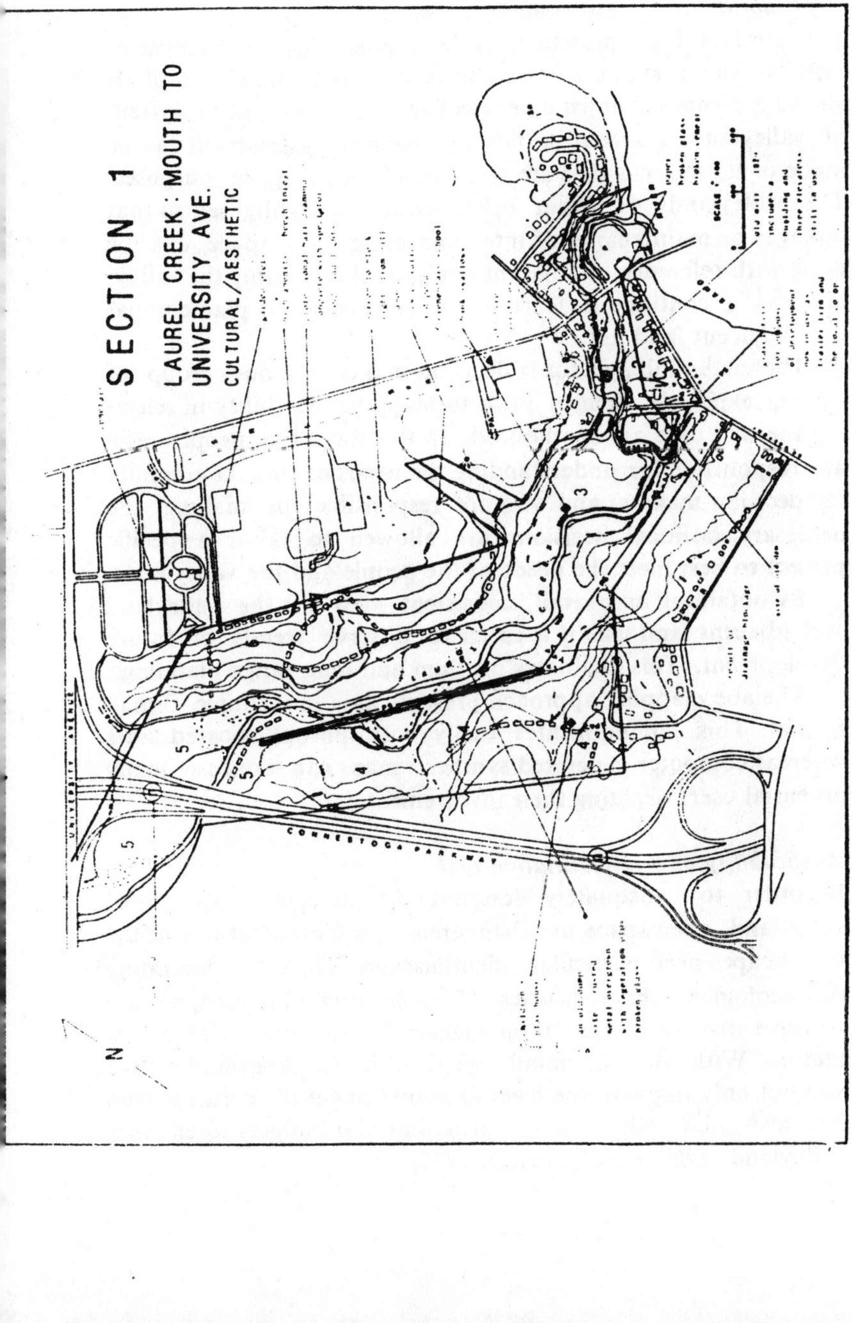

FIG. 9.4 : Representative Section of Cultural Surface of the City of Waterloo

Discussion

Applicability as a decision-making tool

The methodology presented in this paper has a number of aspects which support its value as a decision-making tool. It allows a complete, holistic perspective of a valleyland or system of valleylands so that a fair and balanced assessment can be made of those areas that are suitable for open space purposes. The valleylands as linear open space are highlighted so that linkages or maintenance of interconnecting open space can be done with reference to adjacent land. By illuminating the valleyland as a continuous feature it is much easier to plan compatible adjacent land use.

The methodology affords decision-makers (as one group of users), explicitly defined prior to mapping, flexibility in selecting the way data are synthesized. In this way, the people who are responsible for understanding the needs of the community *i.e.* decision-makers) and who are responsible for making the necessary political decisions are allowed to tailor a specific project to best meet the needs of the people and the valleylands.

By obtaining an overall, consistent view of the valleyland and adjacent land uses it is easier to observe trends of future development. This facilitates medium and long-range planning.

The abc mapping approach presents data in a simple 3-map format. This communicates easily with unsophisticated users whereas computer generated synthesis maps can confuse many potential users negating their involvement.

Applicability as a rehabilitation tool

In order to adequately design and manage a system of valleyland open space in urban areas, the biotic features of the valleyscapes need particular identification. Then by describing the ecological characteristics of each area classified, we can perceive the need for its management and the related cost factors. With this in mind, we can design programmes that will not only improve the level of management of existing land but also plan rehabilitation programs and budgets to enhance valleyland open space (Dorney, 1979).

Cost and time constraints

The useable life of the data displayed varies considerably. The abiotic physical base information is relatively permanent. An exception is water quality, flow rates, and channel characteristics which may vary frequently according to changes made upstream or changes made in adjacent land use. Biotic and aesthetic features will need updating every 5-10 years, depending on the rate of nearby development and plan community dynamics.

The actual time needed to complete the City of Waterloo study was 3 months. The City had 35 miles or 58 km of stream valley. Using a team of three specialists to do the abc mapping at 1 : 5,000 scale and adding a supervisor to the project would make the total effort equivalent to about one-man year. This can be easily translated into cost depending on professional rates of pay. For Ontario in 1980 dollars this would cost $20,000 to $30,000 depending on whether it were done by city staff or contracted out to private consultants.

CONCLUSIONS

The urban valleyland mapping described divides the environment into three components or subsystems—abiotic, biotic and cultural/aesthetic. Production of this 3 map series is technically easy to accomplish, is cost effective, and is easily communicated to various user groups. By concentrating attention on the most natural area remaining in urban areas, *i.e.* the valleylands, nature protection is enhanced along with associated recreational pursuits. By mapping valleylands, their destruction by engineering works will be less likely since abiotic, biotic, cultural/ aesthetic functions and values will be explicitly described.

Acknowledgements

We would like to acknowledge the assistance of the Ontario Ministry of the Environment in providing a grant for this study, and to Mr. Grant Allen, Ms. Barb Cummings, Mr. Don Diebolt, Ms. Marg McKegney and Mr. Doug Platt.

REFERENCES

Adindu, G. and P.F.J. Eagles. 1977. A Manual for Environmentally Sensitive Area Planning and Management. Faculty Env. Studies, Univ. Waterloo, Ontario. 95 pp.

Brady, R.F. et. al. 1979. A Typology for the Urban Ecosystem and its Relationship to Larger Biogeographic Landscape Units. *Urban Ecology*, 4 : 11-28.

Christiansen, E.A. (Editor). 1970. Physical Environment of Saskatoon. National Research Council Public. No. 11378, Ottawa. 68 pp.

Curry-Lindahl, Kai. 1978. Background and Development of International Conservation Organizations and Their Role in the Future. *Environmental Conservation*, 5 : 163-169.

Dansereau, P. and G. Pare. 1977. Ecological Grading and Classification of Land-Occupation and Land-Use Mosaics. *Geographical Paper* No. 58. Fisheries and Environmental Canada, Ottawa.

Dorney, R.S. 1973. Role of Ecologists as Consultants in Urban Planning and Design. *Human Ecology* 1 (3) : 183-200.

Dorney, R.S. 1977a. Planning for Environmental Quality in Canada : Perspectives for the Future. Major Theme Paper presented to the Canadian Institute of Planners ; Annual Meeting, Toronto, Canada, June 27.

Dorney, R.S. 1977b. Biophysical and Cultural-Historic Land Classification and Mapping for Canadian Urban and Urbanizing Landscapes. *Ecological Land Classification Series No. 3*, Environment Canada, Ottawa. Pp. 57-71.

Dorney, R.S. 1979. The Ecology and Management of Disturbed Urban Land. *Landscape Architecture*. Pp. 268-272.

Dorney, R.S. and M.G. George. 1970. An Ecological Analysis of the Waterloo-South Wellington Region. School of Urban and Regional Planning, University of Waterloo, Waterloo, Canada.

Dorney, R.S. and E.S. Higgs. 1978. Urban Valleylands Study of the City of Waterloo. Ontario Ministry of the Environment Experience '78 Programme, Toronto. 298 pp.

Dorney, R.S. and D.W. Hoffman. 1979. Development of Landscape Planning Concepts and Management Strategies for an Urbanizing Agricultural Region. *Landscape Planning*, 6 : 151-177.

Dorney, R.S. and G.B. Priddle. 1973. Study, Inventory, Analysis and Assessment of Urban Environmental Quality: A Case Study of Waterloo, Ontario (Canada). *Faculty of Environmental Studies*, University of Waterloo, Waterloo, Canada.

Fabos, J.G. 1981. Regional Ecosystem Assessment : An Aid for Ecologically Compatible Land Use Planning. In : *Perspectives in Landscape Ecology*, Proc. Intern. Congress Netherlands Soc. Lands. Ecology, Veldhoven. Pp. 163-173.

Francis, G.R. and P.F. Eagles. 1975. A Study of the Environmentally Sensitive Areas for the Environmental Policy Plan of the Regional

Municipality of Waterloo. Department of Man-Environment Studies, University of Waterloo, Waterloo, Canada.

Leopold, L.B. 1969. Landscape Esthetics. Conifer (Pacific Northwest Chapter of the Sierra Club, Seattle, Washington), Vol. 10 (4) : 13-20.

Lewis Jr., P.H. 1967. The Natural Landscape Resources of Wisconsin. Department of Administration, State of Wisconsin, Madison, Wisconsin.

Linton, D.L. 1968. The Assessment of Scenery as a Natural Resource. *Scottish Geographical Magazine* 81 : 221-38.

Maniate, P.M. and D.C. Carter. 1973. The Evaluation of Intangibles in Benefit-Cost Analysis : A General Method. Policy Branch, Environment Canada, Ottawa.

McHarg, I. 1969. Design with Nature. Natural History Press, Garden City, New York. 197 pp.

McIntosh, R.P. 1976. Ecology Since 1900. In : B.J. Taylor and T.J. White (Editors) Issues and Ideas in America. University of Oklahoma Press, Oklahoma. 380 pp.

Sargent, F.O. 1965. Scenery Classification System. Agricultural Economist. Pp. 65-66.

Tokyo, Metropolitan Government. 1974. Map of Actual Vegetation of Tokyo Metropolis (8 sheets).

(5) The budget should be concentrated on transportation for [illegible] investment.

[illegible] new [illegible] and more beautiful Calcutta [illegible] recommendations for immediate action. [illegible] proposed to produce not the city beautiful but the city [illegible]

[illegible] selection of the subject matter for immediate [illegible] development, urban renewal, shelter [illegible] community services and [illegible]

[illegible]

[illegible]

Metropolitan planning is very much similar to regional planning [illegible] based on the [illegible] methodology [illegible] of planning. In these metropolitan [illegible] the end of the continuum [illegible]. [illegible] spatial, temporal and sectoral framework. The spatial perspective of metropolitan planning adopts the macro, the micro and the meso approaches to deal with urban problems in a comprehensive manner. Sectoral and temporal aspects tackle the problems in a more systematic way.

The case studies from the developing countries represent the same problems in every large metropolis. Disintegration of physical and economic planning of the metropolis with the national level is the main problem which hampers the speedy solution of the urban problems. The metropolitan areas are suffering from high density of population, congestion, lack of transportation, housing and environmental pollution. The respective governments are busy to solve these problems [illegible]

TEN

D.R. PHILLIPS

TRADITIONAL AND MODERN HEALTH SERVICES IN NEW PLANNED COMMUNITIES
The Hong Kong Experience

INTRODUCTION

The development of new towns in Hong-Kong

DURING the course of the past decade, Hong Kong housing planners have espoused the concept—widely adopted elsewhere in the world—of providing accommodation in newly created communities, "new towns". However, with typical dynamism, the scale and speed of developments in Hong Kong new town building outrank those found anywhere else in the world and particularly in Western Europe and the United States (see, for example, Osborn and Whittick, 1977 ; Champion, *et al.* 1977). At the moment, there is considerable optimism that these towns will provide healthy new living environments for their inhabitants and that they will, as far as possible, be self-contained in service and employment provision. However, sceptics suggest, perhaps with justification, that although housing units are being provided quickly, service standards adopted are broadly inadequate and will result in long-term insufficient provision in a number of related fields such as health, welfare, education and community facilities. Medical care planning, in particular, is complicated by the fact that traditional Chinese medicine provides care for at least a proportion of the population and Chinese medicine is not explicitly incorporated into Hong Kong Governments' planning. This paper, in part, addresses the question of the role of traditional and western medical care in the new towns. As a result, this topic is of considerable interest to current urban geographers as it considers the question of how very rapid modern urbanisation can come to terms with traditional practices in health and welfare.

During the mid-1970s, hopes were running high that the provision of homes for some two million persons in the three main new towns of Sha Tin, Tuen Mun and Tsuen Wan in the New Territories (NT)—as rural an area as exists in Hong Kong—would solve the housing needs of an increasing population (Pryor, 1978 ; Drakakis-Smith, 1979 ; Phillips, 1980). The population growth has been from natural increase and rapid immigration, whilst housing pressure is added to by squatter clearance and the rehousing of boat dwellers and residents from Mark I, II and III resettlement estates built from the mid-1950s onward. The major influx of illegal and legal immigrants from China during the 1970s and, since mid-1978, the arrival of 'transient' refugees from Vietnam, have added to the Territory's natural increase to result in a net population gain of some three hundred thousand people (enumerated) between mid-1978 and mid-1979. Just what the increase was, until the joint action by the Chinese and Hong Kong authorities in late 1980 to clamp down on illegal immigration, and what the accuracy of the April 1981 Census is, remain a matter of considerable speculation amongst professionals in the Government and the private sector. The 1981 Census figure of 5,021,066 is actually below the Government's own end-of-year estimate for 1980 (which was 5,147,900), and considerably below the estimate for the end of 1981 (5,207,000), which does not encourage confidence in the result.

Whatever the population increase, just as important is the growth of 'nuclear' families, with two or three children, who want self-contained accommodation without overcrowded, shared facilities. The increasing affluence of Hong Kong's labour force means that there is a pent-up demand for good quality accommodation. In terms of housing, it is evident that even the very considerable number of 30,000-35,000 units being provided annually by the Housing Authority (mainly in the new towns) will be totally inadequate to the task of meeting the

Government's expressed aim of providing a self-contained home for every family. Sheer physical constraints such as bottlenecks in the construction industry and in transport, to say nothing of financial costs, make adequate provision in terms of numbers of houses very problematic. However, this need is made all the more pressing by excessive land prices, which reached record levels in the summer of 1981, although settling back in 1982, and the still extremely high rents and housing prices in the private sector of Hong Kong's housing market. As a result, Hong Kong planners have talked for some time of the concurrent expansion of several existing townships in the New Territories as actually being a phase of additional new towns, at Tai Po, Yuen Long, Fanling-Sheung Shui, with a completely new development at Junk Bay planned for the mid-1980s (see Figure 10.1). The development programme for the last-named settlement was produced in September 1981. The planned populations of the new towns are enormous in anybody's estimation : Sha Tin and Tuen Mun will each house over half a million persons and Tsuen Wan one million by the late '80s and figures for the other settlements are under constant upward revision.

Planning health and welfare services

Obviously, such enormous developments present many related problems in the provision of social services in general and health, welfare, educational and community facilities in particular. The overwhelming need for more and more housing units is, it seems, being used by some as an excuse for abrogating the responsibility of providing full adequate levels of services in the new communities, under the blanket apology of 'insufficient funds'. This means that, in the long run, there is the very real danger of these new towns becoming second-class satellite concrete jungles with little in the way of community spirit, and with degenerating physical appearance as years progress. Hong Kong Government is often referred to as a form of 'benevolent dictatorship', yet it relies heavily upon government-subvented voluntary agencies to provide a large proportion of social welfare services in existing settlements and is doing the same in the new settlement. However, these agencies often depend upon a close contact and cooperation developing within local communi-

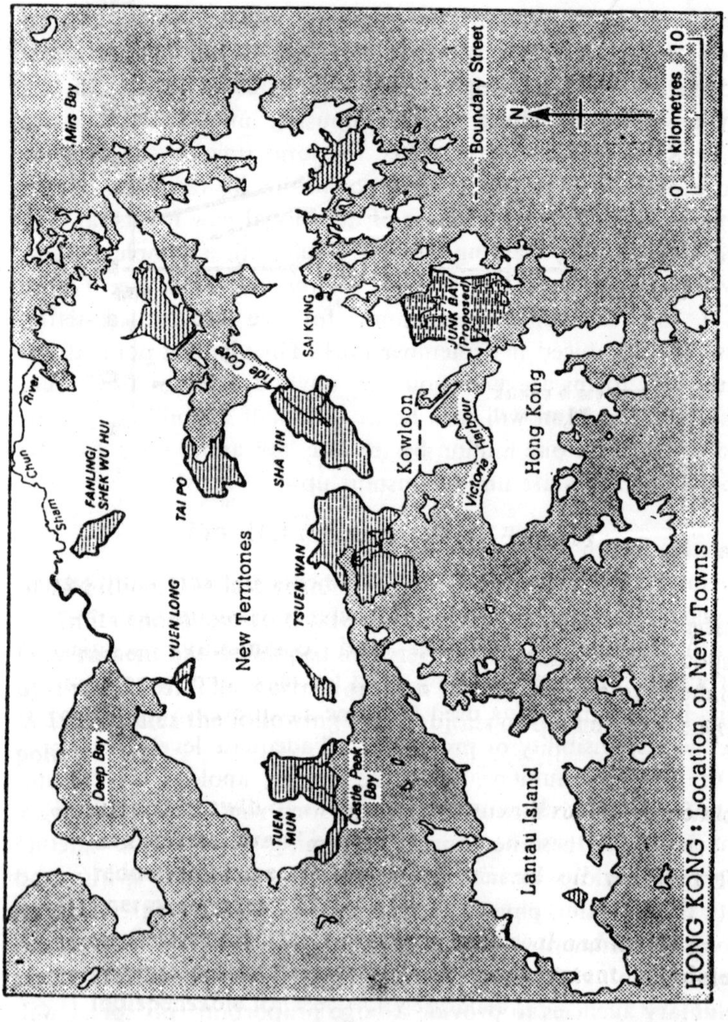

FIG. 10.1 : *Hong Kong : Location of New Towns*

ties and there may be some question as to how successfully (or how quickly, which is as important), these voluntary agencies will be able to expand and transplant their functions to the newly urbanised areas.

In Hong Kong as a whole, comparatively low standards of housing and welfare provision are defended upon pragmatic grounds—that it is impossible to provide Western European (or even some East Asian) space standards in the crowded local conditions, and that 35 square feet of living space per person in a properly constructed Housing Authority block is preferable to a tin shack on the hillside. As Lee (1981) writes, squatting is still a major problem both in human terms and politically. As many as three-quarters of a million squatters still cling tenaciously to increasingly perilous slopes and the ever-present risks of fire and landslides remain and have become more acute and electricity and other services have become even more overloaded.

A key document in the developments is the more or less 'restricted' Hong Kong Outline Plan (HKOP), retitled in 1982 Hong Kong Planning Standards and Guidelines (HKPSG). Originally proposed in 1965, updated in 1974, and under continuous review, it defines 'acceptable' development standards and planning strategies in the light of existing government policies, even if real strategic forward planning is very thinly represented in this document. Amongst other things, HKOP specifies the nature and range of community facilities (including welfare facilities) which are officially deemed to be acceptable and necessary and with which developers must conform. It is the heavy reliance upon the norms laid down by HKOP (adopted for speed and simplicity by new town planners so that 'off the peg' services can be plugged into each new development) that cause some concern to those with greater vision than average amongst local planners.

HKOP (HKPSG) provides the 'nuts and bolts' of planning norms and, arguably, a rather inflexible system results. Many services are planned purely on a *per capita* population basis, for example, primary and secondary schools, some types of medical clinics, community halls, estate community centres and police stations. Others are non-population based in public housing estates, such as hostel accommodation, sheltered workshops and rehabilitation centres. If an apparently egalitarian system from estate to estate does develop as a result, it may often be inequitable as different estates with different population structures will, over time, require varied types of service provision.

HKOP (HKPSG) naturally enough in the crowded circumstances of Hong Kong, places great emphasis upon physical size of services (square metreage) and, as Table 10.1 illustrates, standards are certainly based upon Hong Kong rather than international norms, with schools running bi-sessionally to maximise use and use-rates of other welfare services exceeding what would be normally expected in an industrialised country. The reviews of the planning guidelines make it possible to incorporate minor revisions from time to time, usually aimed at improving living standards for residents. Recent suggestions regarding medical facilities are now upgrading provision although, at first glance, some primary care facilities seem actually to be being reduced. For example, the provision of private clinics (one private general practitioner doctor per clinic) is now to be 1 : 7,500-10,000 residents. This appears quite unfavourable in comparison with provision of such general practitioner services in developed countries where 1 : 2,000-3,000 persons might more commonly be expected. However, additional government clinics will provide virtually free care nearby to most public housing estates either in clinics, health centres or polyclinics. These will support the private practitioners who comprised one of the main groups wishing to adjust this standard to enable them to attain enough patients to earn sufficient income from fees.

The residents for whom publicly-provided services are of paramount importance are, of course, the less well-off new town residents, living in public housing estates. Some 70 per cent of residents in most of the new towns will live in public housing—those in private housing and home ownership schemes are somewhat better placed financially to 'supplement' public services with privately bought ones, if necessary. So, the HKOP (HKPSG) standards in the end affect mostly the financially-poorer two-thirds of residents in the new towns.

If it is assumed that these basic norms are, in fact, provided (and there is evidence that this is not the case as high rents in some of the public housing estates are resulting in a low rate

TABLE 10.1 : **Selected Hong Kong outline plan (Hong Kong planning standards and guidelines) standards for "community facilities"**

(Government and privately-provided facilities) *Medical Facilities* (based upon White Paper recommendations)		*HKOP Standard**	*HKPSG (1982 suggestions)*
Hospital beds		target 5.5 beds/1,000 persons	similar, but more flexible distribution of beds amongst specialities
Private Clinics (in public housing estates)		1 doctor/6,000 persons	1 doctor/7,500-10,000 persons to be provided in public housing estate where no purpose built govt. clinic available/planned in vicinity
Government Clinics (in public housing estates)			
Health centres/polyclinics		Flexible	
Education facilities	serving area of—	*Bisessionally-used classrooms*	
Kindergarten (not statutory)	0.4 km radius	1 classroom (90 places)/2,700 persons	unchanged
Primary	0.4 km radius	1 CR (90 places)/850 persons	unchanged
Secondary	3.2 km radius	1 CR (unisessional, 30 places)/520 persons	unchanged
Special schools			
Higher Education		not specified	

TABLE 10.1 (*Contd.*)

(Government and privately-provided facilities) *Medical Facilities* (based upon White Paper recommendations)	*HKOP Standard**	*HKPSG (1982 suggestions)*
Community facilities		
Child care centres/kindergarten	No set standard : Social Welfare Dept. Suggests 100 places/20,000	
Youth centre	1 centre/20,000 persons	unchanged
Children's centre	1 centre/20,000 persons	unchanged
Estate community hall	1 hall/20-50,000 persons	unchanged
Estate community centre	1 centre/50.000 persons	unchanged
Community centre	1 centre/100,000 persons (where no estate facilities exist)	unchanged
Social centres for elderly	1 centre/30,000 persons	

Note : Figures are for places per total population.

of take-up of sites by certain types of services), it is still arguable that these new towns are not the panacea for housing and social problems. One example to support this contention is that the Government White Paper of 1974, which underlies much thinking on the development of medical and health services in Hong Kong, lays emphasis on improving general standards by increasing provision and building up medical and health services in the new towns and New Territories in general. At present, Hong Kong hospital beds are provided at something like 4.4 beds per 1,000 persons, planned to reach a target of 5.5 beds per thousand in the long term. Even if allowance is made for the methods of calculation used, this gives a very unfavourable international comparison (U.K., for example, provides about 10.1 beds per 1,000 ; USA about 8.9 beds per 1,000 and USSR about 9.6). All that the 1982 review will mean is that more flexible allocation of hospital beds amongst specialties will be possible.

The achievement of even a modest figure of 5.5 beds per 1,000 for the New Territories is being hampered by the need to provide totally new facilities (such as the 1982 new teaching hospital in Sha Tin new town) and the difficulty of recruiting medical staff for the New Territories. The government is still relying considerably upon the voluntary sector here—and in this faces real problems of providing adequate quality in medical facilities. For example, Pok Oi subvented district hospital at Yuen Long (one of the expanded towns recently rechristened a 'new town') relies upon half-a-dozen Burmese-trained medical officers, with a medical superintendent 'on loan' from Hong Kong government. The hospital would regard it as a considerable achievement if it were able to attract in Hong Kong University medical graduate. Similarly, public housing estate primary care doctors can see between 50 and 60 patients per day, and, in an existing New Territory polyclinic, each medical officer is seeing 150-plus patients in an eight hour day—hardly encouraging to high standards of professional judgement.

Even in the newly-opened Lek Yuen Health Centre in Sha Tin, each government medical officer sees some 120 patients per day. To balance this, charges at government clinics throughout Hong Kong are moderate (and free for the genuinely indigent), but numbers using facilities are such as to make non-urgent clinic visits most undersirable for patients. Queues can start as early as 4 a.m.—the elderly may regard queuing as a form of social activity but it is not conducive to good preventive medicine which relies on swift and easy access. 10 a.m. sees queues beginning in clinics for tickets for afternoon session. Due to the need for a profit-making outlook by medical practitioners in general practice in the new towns to pay for rental and to earn income, the Estate Doctors' association feel that the norm of 1 Doctor per 6,000 was too generous and that 1 per 7,500-10,000 should enable their members to make a better living.

Due to high rates of utilisation, overall standards are thus not even as good as official norms suggest. In the medical field, the spectrum of diseases needing treatment in Hong Kong is changing, following the pattern of other developed nations. There are more chronic conditions and fewer acute infectious cases, and more old people provide juvenile Hong Kong with a problem which it cannot really comprehend or readily solve. There are very few consultant geriatricians in the government and subvented sectors. In the welfare field, community facilities for all age groups are grossly overcrowded and in constant use wherever possible. The proportion of Hong Kong's population aged over 65 years rose from 4.5 per cent in 1971 to 6.6 per cent in 1981. This increase is likely to be maintained as the previously very young population of the Territory ages and as birth rates steadily fall (Hong Kong Government 1982). Therefore, this is one sector of the medical and welfare field to which attention must be directed. The new towns may not currently house a large number of elderly but the future will see many of their residents falling into this category. Whether the challenge of meeting the medical and welfare needs of a gradually ageing population can be met with the standards of provision at present planned may be questioned. However, there is another form of medicine used here, traditional Chinese medical practitioners and this paper now addresses the possibility of including this sector into the official medical planning.

Traditional medicine's contribution to health services in the new towns

It is fitting to question whether the KHPSG norms are as

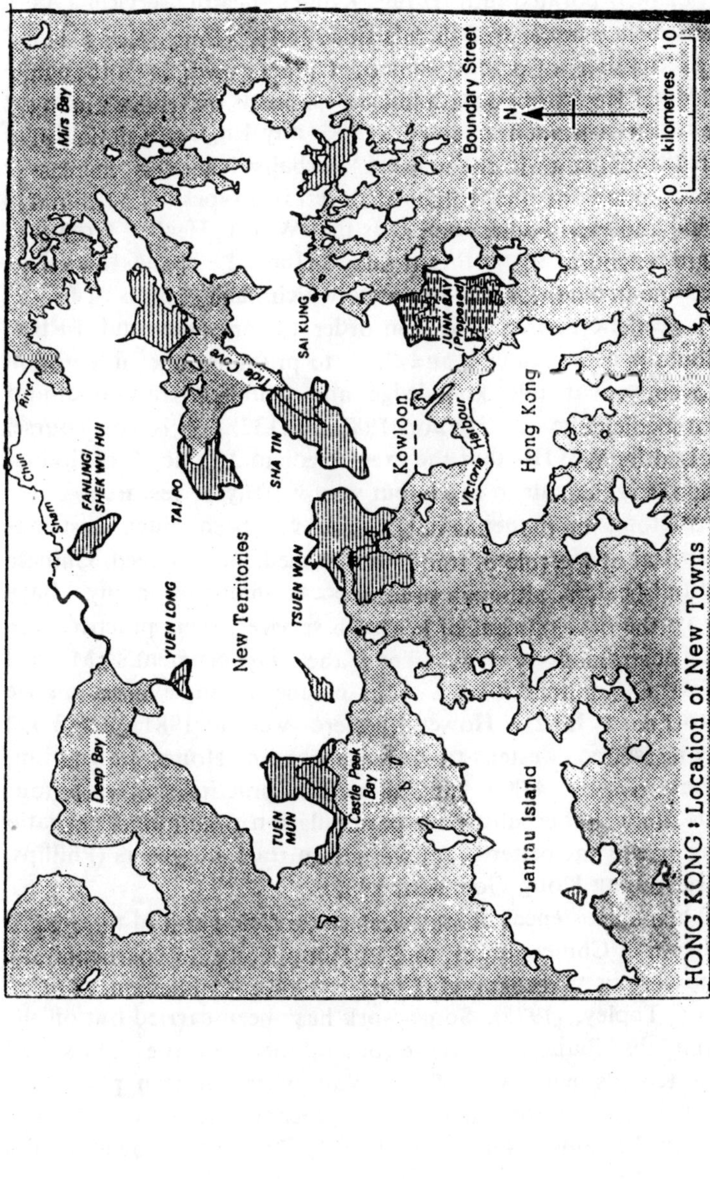

FIG. 10.2 : *Hong Kong : Location of New Towns*

inadequate as it at first appears. It was suggested earlier that Hong Kong, as many other countries at lesser stages of development, has a considerable input from traditional ('non-western') medical practitioners. However, these are not officially planned for, or incorporated into, Hong Kong's health services. As a Chinese society under British rule since 1841, Hong Kong has a large number of practitioners of Chinese medicine although the Medical Registration Ordinance prevents them from claiming to be western-trained doctors and any English translation of their title must contain the word "herbalist". There is increasing recognition of the value of certain types of traditional medicine and even bodies such as the World Health Organisation are encouraging this attitude. Indeed, it has a current programme in traditional medicine which has the aims of fostering a realistic approach to it in order to promote and further contribute to health care and ". . . to promote the integration of proven, valuable knowledge and skills in traditional and western medicine . . ." (Vuori 1982, p. 132). It is, of course, recognised by WHO that the exact medicinal value of traditional medicine is difficult to assess but very worthy of research.

In Hong Kong there has not, however, been much systematic appraisal of the role of traditional medicine. Indeed, Chinese traditional healers, although used occasionally by many, have no say in the development of health services, their practices are much constrained by law, and the Government's Medical Advisory Committee does not include a single Chinese-style doctor (Lee 1982). However, there were in 1981 some 3,900 fully-registered western-trained doctors in Hong Kong but, probably, over 4,000 Chinese medical practitioners of various kinds (mainly herbalists), their population : practitioner ratio being somewhat better than for western-trained doctors (Phillips, 1981a ; Hong Kong Government 1982).

The co-existence of both western medicine and traditional medicine in Chinese cultures, and in Hong Kong in particular, is now very well recognised (Crozier, 1968 ; Kleimeman, *et al.* 1975 ; Topley, 1975). Some work has been carried out on the spatial distribution of these practitioners in the 'oldest' of Hong Kong's new towns, Tsuen Wan, where western physicians and herbalists were found to be located quite near to each other in the same sub-districts (Lee, 1972). In this town at the time of the survey, (1971), 92 physicians and 111 herbalists were found (there would therefore be more traditional practioners since acupuncturists and bonesetters, which together comprise about 30 per cent of Chinese practitioners, were not included) and this supports the suggestions above that, in numerical terms, traditional medical practitioners are more important than western trained (Lee, 1974). This argument, of course, really only holds good for primary health care, since few, if any, traditional practitioners could cope with hospital-type cases. Nevertheless, it seems that the Chinese sector of medicine is very important numerically in Hong Kong.

However, in spite of this, there is little knowledge of the relative levels of dependence on western and Chinese practitioners upon which Hong Kong planners could base judgements. This seems to be an important weakness since Hong Kong's planning norms appear to be placing an implicit reliance on the Chinese medical sector and are therefore planning western facilities (primary care in particular) to lower levels than might otherwise be expected. Lee (1972) suggested that respondents in a survey were rather more likely to utilise Western than Chinese medical care, but this work has not been much extended to the latest of the new towns. With greater utilisation of Western medicine and its control over the social organisation, planning and technical content of medicine in Hong Kong, the dual systems of medical care can be characterised as Western medical dominated (Lee 1974), but the extent to which this can be allowed for in planning is not fully clear.

In the rapidly growing and modernising society of Hong Kong, it seems important that more information is obtained about utilisation patterns. Such information will form essential planning data and, recently, some research has been conducted to attempt to increase knowledge of the behaviour of new town residents with regard to health care utilisation.

Sha Tin is one new town which is fast developing, with a current population of over 150,000 and a planned population of some 560,000 by 1987-1988, rising to 750,000 in the 1990s. A well-established social science research design was employed, in which housing areas of known social composition with access to a known range of facilities were selected, so that the 'consumer' behaviour of residents could be investigated (Thomas, 1976 ; Phillips, 1981b). Interviews were conducted on two occasions, in July-August 1981 and July-August 1982, on each of which 200 households were visited and their utilisation of traditional and western health care facilities discussed. Respondents were asked to indicate how often they used facilities (Chinese and western), which facilities attended, and how long they had been using any particular facility. They were also asked their attitudes to, levels of satisfaction with, and confidence in, various medical services. The results from the 1982 survey are at present only available in preliminary form but both surveys indicated considerable degree of reliance on chinese medical practitioners, although western medicine did dominate usage.

Proportions of respondents (1981) usually or always attending Western doctors for primary care varied from 74 per cent in public housing to just over 80 per cent in privately-owned housing. When children were considered, proportions were somewhat higher and over 80 per cent usually attended Western doctors. However, the converse is that, for primary care, between 20 and 25 per cent of respondents were relying mainly on Chinese style practitioners.

Preliminary analysis of the 1982 survey supports these results. Dependence on traditional practitioners was somewhat higher in the lower social status public housing, where up to 44 per cent of respondents reported attending a herbalist at least once for medical consultations during the twelve months previously. In higher status home ownership schemes, this proportion fell to 20 per cent. Utilisation of traditional medical services was much higher amongst female respondents and, to some extent, mothers would take their children to such practitioners, depending upon the nature of the medical problem to be examined. As a second stage of the survey, the herbalist practitioners in the estates were interviewed and, by and large, these confirmed that women and children made up the majority of their patients, men usually only comprising between 20 to 40 per cent of attenders.

Therefore, the results briefly reported here do support the

contention that traditional medicine (herbalists only, as acupuncturists and bonesetters are not included in the above discussion) do provide a valuable and quite popular service, even if not used by a majority of residents in the new town. Admittedly rates of dependence on traditional medicine are not as high as in many other countries, where in some areas, almost the entire population is served by traditional practitioners (see, for discussion, Ramesh and Hyma, 1981 ; Maclean and Bannerman 1982). However, the significance of the numbers reporting attendance is such that they should in some manner be incorporated into the official planning process for medical services. At least a more thorough investigation is suggested so that the precise nature of their contribution can be calculated.

CONCLUSIONS

This paper has considered the nature of planning of an essential urban service namely, medical care, in a new town setting in a Territory where the co-existence of modern and traditional medical systems is a complicating factor. It may be accepted that some easily implementable guidelines are essential for the provision of services in times of rapid planned urbanisation, but it is possible to call into question some of the norms adopted in the Hong Kong new towns for medical and welfare facilities. In particular, this paper suggests that traditional Chinese medical facilities should be explicitly included into the planning process in order to provide a more realistic picture of the services actually used by residents. The fact that traditional medical practitioners shops are to be found in some numbers in the new towns indicates that they are financially viable and supported by considerable numbers of paying patients. However, until their precise contribution is established, the extent of any shortfall or deficiency which may (or may not) exist in the Hong Kong planning norms cannot be determined.

The urban geographer does have an important role to play in monitoring service standards adopted, particularly in highly-planned urban developments. This role can, perhaps, be regarded as one of commenting on standards in the hope that they will be maintained and of highlighting areas for improvements, in order to enhance the quality of life for residents in such developments. With their expertise in survey design and an understanding of a 'welfare approach' in human geography as advocated by Smith (1977), urban geographers may contribute to a monitoring of standards in urban growth. If they are to do more than this, their work must be seen to be of value by professional planners and service providers and it is to this end, liaising and assisting these professionals, that it seems sensible to aspire.

To be fair, Hong Kong is actively facing up to some of its problems. The provision of social and welfare facilities is only one aspect of the struggle to provide basic accommodation units and infrastructure. It is only to be hoped that higher standards will be aimed for in the future if the new town schemes are not to lay up great problems for the future, greater, perhaps, than any successfully met so far. These schemes are visionary and visionary new standards should also be adopted. Now is not the time for a reliance on inadequate, insufficient old standards and these developments provide a great opportunity to give these new communities the best aspects of western and Chinese health care. The current means do not justify the end and the creation of potential satellite slums will not be the way forward. Unfortunately, the scale of the entire problem provides a numbers game *par excellence*.

Acknowledgements

The author gratefully acknowledge the support of The Nuffield Foundation for financing the research upon which this paper is based. In addition, he is grateful to D.E.K.Y. Chen, Director, The Centre of Asian Studies, The University of Hong Kong, for providing there an academic base for the project during its fieldwork stages in 1981 and 1982.

REFERENCES

Champion, A.G., Clegg, K. and Davies, R.L., 1977, *Facts about the new towns*, Retailing and Planning Associates, Corbridge.

Crozier, R.C., 1968, *Traditional medicine in modern China*, Harvard University Press, Cambridge, Mass.

Drakakis-Smith, D., 1979, *High Society*, Centre of Asian Studies Occasional Papers and Monographs No. 40, Centre of Asian Studies, The University of Hong Kong.

Hong Kong Government, 1982, *Hong Kong 1982*, Hong Kong Government, Hong Kong.

Kleineman, A., Kunstadter, P., Alexander, E.R. and Gale, J.J., eds., 1975, *Medicine in Chinese Cultures*, U.S. Department of Health, Education and Welfare, Public Health Service, National Institute of Health Publication, NIH 75-653.

Lee, M., 1981, 'A squatter boom', *Far Eastern Economic Review*, 28 August, 34.

Lee, R.P.L., 1972, *Spatial distribution of modern Western and Traditional Chinese medical practitioners in an industrialising Chinese town*, Social Research Centre Monograph, The Chinese University of Hong Kong.

Lee, R.P.L., 1974, *The stratification between modern and traditional professions, a study of health services in Hong Kong*, Social Research Centre Monograph, The Chinese University of Hong Kong.

Lee, R.P.L., 1982, 'Comparative studies of health care systems', *Social Science and Medicine*, 16, 629 612.

Maclean, U. and Bannerman, R.H., 1982, 'Utilisation of indigenous healers in national health delivery systems', *Social Science and Medicine* 16, 1815-1816.

Osborn, F.J. and Whittick, A., 1977, *New towns : Their origins, achievements and progress*, Leonard Hill, London ; Routledge and Kegan Paul, Boston.

Phillips, D.R., 1980, 'New towns bring new hope for expanding Hong Kong', *Geographical Magazine*, LIII, 180-187.

Phillips, D.R., 1981a, 'Socio-economic factors and the planning of medical facilities in new communities : a case-study in Sha Tin new town', paper presented to the Commonwealth Association of Planners Asian Regional Conference, The University of Hong Kong, December.

Phillips, D.R., 1981b, *Contemporary issues in the geography of health care*, Geo Books, Norwich.

Pryor, E.G., 1978, 'Redevelopment and new towns in Hong Kong', in L.S.K. Wong, ed., *Housing in Hong Kong : a Multidisciplinary study*, Heinemann, Hong Kong, 266-286.

Ramesh, A. and Hyma, B., 1981, 'Traditional Indian medicine in practice in an Indian metropolitan city', *Social Science and Medicine*, 15D, 69-81.

Smith, D., 1977, *Human geography : a welfare approach*, Edward Arnold, London.

Thomas, C.J., 1976, 'Sociospatial differentiation and the use of services', in D.T. Herbert and R.J. Johnston, eds., *Social areas in cities, Vol. 2*, Wiley, London, 17-63.

Topley, M., 1975, 'Chinese and Western medicine in Hong Kong : some social and cultural determinants of variation, interactions and change', in A. Kleineman *et al.*, eds., *Medicine in Chinese Cultures*, U.S. Department of Health, Education and Welfare, 241-271.

Vuori, H., 1982, 'The World Health Organisation and traditional medicine', *Community Medicine*, 4, 129-37.

ELEVEN

J. GORDON ADAMS

POST-WAR PLANNING IN THE GLASGOW REGION

HISTORICAL INTRODUCTION

As WITH all other cities, the present problems of Glasgow need to be considered in a historical context. Although Glasgow has a Cathedral which was founded in 1124, and a University which was founded in 1450, it is as a city of the industrial revolution that Glasgow acquired world-wide fame and notoriety. Glasgow was the largest town in the West of Scotland before the political Union of Scotland and England in 1707, and although, as in most of Scotland, there were violent protests against the Union, it seems clear that Glasgow gained substantially from legal access to English colonial markets in the West Indies and North America during the eighteenth century. Glasgow firms were successful in the sugar, cotton, and, particularly, the tobacco trade, and this success may be reflected in the antipathy in Glasgow to the Jacobite rebellion of 1745. By the second half of the eighteenth century commercial experience and capital accumulation had proceeded sufficiently that the loss of the tobacco trade following the American War of Independence seems to have been overcome with little difficulty.

Transport improvements in Scotland in the late eighteenth century (e.g. the Forth-Clyde Canal of 1790) were both a response to earlier development and a means of further expansion. The cotton industry increased enormously between 1780 and 1835 and provided the basis for the expansion of population in Glasgow (77,000 in 1801, 256,000 in 1841). The textilé industry acted as a catalyst for other enterprises (particularly chemicals and textile engineering), and expertise in metal engineering was stimulated by a massive expansion in the iron industry (based on local coal and iron) from about 1830.

The increase in population in Glasgow in the early nineteenth century, however, needs to be seen both as a response to increased demand for labour in the cities, but also as a response to increasing misery in the countryside, particularly in Ireland and the Scotish Highlands and Islands. Rural emigration from these areas had occurred in the eighteenth century, but after the Napoleonic Wars the equilibrium of the economy seems to have collapsed. Increasing population combined with declining incomes and coercive population removal, so that huge numbers of poor, and sometimes starving, migrants moved both to the expanding British cities and to North America. By mid-century perhaps one-third of the population of Glasgow was of Irish origin-although at least half of the Irish migrants may have been Protestants, contrary to local myth.

Conditions in Glasgow in the mid-nineteenth century horrified outside observers. Overcrowding and wretched housing produced astounding health problems with epidemics of typhus and cholera, along with serious political disturbances (e.g. 1816). Both private and public authorities did react to tackle the problems although initially with very limited success. The town of New Lanark was developed in the early nineteenth century as an experiment in paternalistic utopianism. In the second half of the century a series of public Acts attempted to tackle the worst problems of health and over-crowding. With regard to the Glasgow public health measures Marx concluded that "the alarmed bourgeoisie had recourse to municipal action", but humanity may also have been responsible, as well as fear.

The boom years of the iron industry in the Glasgow region were from 1830 to 1870. The transfer from an iron industry to a steel industry was not easily made but massive expansion did take place between 1880 and 1890 and accompanied amazing growth in ship-building. In the peak year of 1913 the Clyde launched 18 per cent of total world output of shipping, a total greater than either Germany or the United States (Slaven, p. 178).

By the first world war, then, the Glasgow area had a flourishing economy apparently based on technologically advanced industries with linked engineering enterprises literally of all kinds. The motor-car and aircraft industry were established in the area just before the war. In retrospect the beginnings of economic decline can be traced to the nineteenth century when local industries were no longer world technological leaders, but the speed of decline after the first war does seem to have astonished most observers. The ship-building industry may have become over-dependent on naval work ; local iron and coal supplies lost their cost advantage ; the loss of export markets during the inter-war period was particularly severe for industries with a very limited home market ; but whatever the reasons, the Glasgow economy did not make a successful transition from the age of steam, and this is in contrast to the earlier relatively smooth transitions from tobacco to textiles to iron and to steel. Over half the male labour force was unemployed in some of the towns in the Glasgow region in the early 1930's and in spite of tentative regional development measures, unemployment did not improve substantially until the re-armament programme before the second world war.

Post-war physical planning

In 1946 a Clyde Valley Regional Plan was published and many aspects of the present physical structure of the Glasgow region stem from decisions made in that plan. The Clyde Valley Plan also needs to be seen in its historical context. One of the principal authors was Sir Patrick Abercrombie, who had become famous for his efforts to control the sprawl of London and the encroachment of rural land. The process adopted in London was the imposition of a 'green belt' where industrial and housing development was forbidden, along with the building of a series of satellite settlements (New Towns) to cope with population moving out from congested conditions in London. The London green belt was designed primarily for recreational purposes and the English New Towns were the culmination of a process of anti-urbanism familier in British planning theory.

The Clyde Valley Plan was also produced in a period of optimism about the possibilities of 'socialist' planning. War-time direction of the economy had eliminated unemployment and it seemed possible that the national single-mindedness which had been apparent during the war could be harnessed towards 'winning the peace'. Thus there was the apparently astonishing rejection of Churchill's party before the end of war. The new Labour administration introduced a series of social welfare measures (in particular the creation of the National Health Service) and the major industries of coal, railways, electricity and steel were nationalised between 1947 and 1949.

The Clyde Valley Plan was concerned primarily with the twin problems of congestion and housing and was a 'physical' rather than an 'economic' plan. Provision was made for increased land for industry but the plan provided remarkably little emphasis on employment, particularly when the problems of Glasgow in the 1930's are considered. In the event, however, unemployment did not re-appear as a major problem in Clydeside until well after the war. The Clyde Valley Plan proposed that Glasgow should be constrained by a green belt, although preservation of agricultural land was used as the main justification for this policy, rather than provision of recreational facilities as in London. Glasgow in fact has magnificent natural recreational facilities in the Loch Lomond area (about 50 kms. away) and further north. Given the problems of congestion in Glasgow at the end of the war, (a population of 1.1 million in 1951 with population densities of over 1,750 per hectare), a green belt policy meant that large-scale emigration also had to be part of the plan. This was to be achieved partly by the building of New Towns (four were originally planned), although it is likely that Abercrombie would have approved of New Towns as part of the solution for Glasgow congestion, even wlthout a green belt. The City of Glasgow originally opposed the New Town concept and proposed instead that decongestion could be achieved by relatively high-density peripheral development on what was proposed as green belt. Central government, however, through the Scottish Office, was in favour of Abercrombie's proposals and East Kilbride New Town was designated in 1947. Although there was general agreement about the objectives of the Clyde Valley Plan (decongestion and improved housing) it is surprising, in retrospect at least, that a key feature of the plan for the physical regeneration of Glasgow had to be imposed against the wishes of the local elected representatives.

By 1952, however, the Glasgow attitude to New Towns had changed and Cumbernauld was designated as a second New Town in 1956. In the late 1950's plans were also made for 'overspill' movement, i.e., assisted migration to other towns in Scotland, for large scale house-building in peripheral local authority estates, and for 'Comprehensive Redevelopment' (i.e. demolition and rebuilding) in large sites throughout the city.

For almost 25 years after the war Glasgow suffered from an acute shortage of housing. Although the city had experienced very little bomb damage, the post-war over-crowding in houses largely built in the nineteenth century meant that house building became the paramount objective for politicians at the local level. In addition there was a very understandable reaction against the squalor of housing conditions before the war, but in some cases this resulted in approval for demolition where careful rehabilitation might have been possible. In the event the Comprehensive Redevelopment schemes often proceeded very slowly and there are still large vacant sites throughout the city where redevelopment is scheduled but has not yet occurred.

Retention of population in the Glasgow area has now become a stated political goal. This is perhaps surprising in that the smaller population for the city which was the objective after the war (650,000) has not yet been reached. The objective of population retention can be attributed at least partly to simple desire for power and prestige. Partly, however, population retention is wanted because population decline has multiplier effects just as has population increase. Estimates have been made that one extra job is lost following an original decline of between 4 and 7 jobs. Thus it becomes very difficult to control population decline as thresholds are passed and economies of scale are lost. If emigration from Glasgow is a function of both the push from Glasgow and the pull of other areas it is likely that emigration will decline as the difference between unemployment in Glasgow and unemployment in other parts of Britain declines (see Table 5 in the Appendix). The objective of population retention, then, could be partly achieved as a side-effect of a general deterioration in the British economy.

Regional policy

Regional policy in Britain in usually said to have started in the 1930's with some efforts to assist emigration from areas of high unemployment, and with the creation of 'Special Areas' where government could provide infrastructure and loans to businesses. The re-emergence of regional policy after the war should be seen as part of the general move towards greater planning mentioned earlier. The Distribution of Industry Act of 1945 provided for assistance to industry in 'Development Areas'—one of which was Central Scotland. In addition curbs on further development in certain areas (primarily south-east England) were imposed through the operation of a system of Industrial Development Certificates.

In the Glasgow area after the war a large factory-building programme was undertaken by government, building on a pre-war strategy of 'industrial estates'. This programme appears to have been quite successful and a number of international (particularly American) firms were attracted to the area (e.g. I.B.M., Honeywell). In addition the Glasgow area gained from a government policy of 'steering' new private investment to acceptable locations. East Kilbride was particularly successful in gaining new employment in this way (e.g. Rolls Royce and, later, government offices).

Unemployment did not re-appear as a major problem in the Glasgow area during the 1950's as had been feared, although this was at least partly because of high net out-migration. Again, in retrospect, the features of a poor industrial structure (i.e. a high proportion of industries that were declining nationally) along with low productivity were obscured. Conditions did worsen around 1960 and one result was a greater government emphasis on regional policy. In 1963 an important White Paper on Central Scotland (Cmnd. 2188) placed great store on the growth centre concept which had gained academic popularity in Britain following its original formulation in France. A series of growth centres was identified in the Glasgow area and the 1963 White Paper is generally seen as an indication of a change from emphasis on areas with economic problems to an emphasis on areas with potential for economic growth. The identified growth areas were to be stimulated primarily by infrastructure development (particularly transport), but the size of growth area chosen seems to have been too small. In addition relatively little was done to ensure industrial expansion (largely under the control of private industry) as opposed to infrastructure development (largely under the control of the public authorities). The establishment of a car-making plant at Linwood near Paisley was a valiant attempt to re-create a Scottish car industry and here at least was a growth centre according to the French example of 'industrie-clef'. The Linwood plant, however, has never been completely successful, and in less than 20 years has been transferred from the Rootes Group to Chrysler and now to Talbot.

It is difficult to remember only 15 years later that the proposals for growth centres in the West of Scotland in the early 1970's should be seen in the context of a "National Plan" which foresaw that British economic development could be constrained by a national shortage of labour. Irvine was designated as a third New Town in the West of Scotland in 1966, and the designation of Irvine should be seen both as evidence of continued faith in the concept of relatively small growth centres and a continuation of the policy of "thinning out" of the population of central Glasgow. For a variety of reasons, however, substantial development did not get under way in Irvine until the 1970's.

Economic conditions in the Glasgow region worsened appreciably in the late 1960's and by 1971 male unemployment in West Central Scotland was over 9 per cent. In 1974 the West Central Scotland Plan was published by a joint committee of local authorities and central government. As with the earlier Clyde Valley Plan this was a series of proposals rather than an adopted plan, although the 1974 Plan had a much greater emphasis on economic, as opposed to physical, planning. One of the most important recommendations of the 1974 Plan concerned the establishment of a public corporation to assist industry—the proposed Strathclyde Economic Development Corporation (SEDCOR).

Physical and economic planning from 1975 to 1980

Some very important changes in the system of physical and economic planning in the Glasgow region have taken place since 1975. In May 1975 a new structure of local government was introduced and a system of Regional and District local authorities was introduced ; the main island areas have a different system. Strathclyde Regional Council was created as by far the largest Scottish region in terms of population (about 2.5 million). The region includes the traditional Scottish industrial heart-land of the Clyde Valley but also includes substantial rural areas in the north-west and south-east. The City of Glasgow District Council has responsibility for a re-defined Glasgow with a population in 1975 of about 900,000. Regions and Districts do not have completely separate functions—industrial development, for example, is within the jurisdiction of both authorities, but in the present context the key difference is probably that the Regional Councils are responsible for strategic planning, while the District Councils are responsible for housing.

The very laudable attempt, then, to produce local government on a city-region basis has been diluted by the political necessity to keep housing under the control of local (as opposed to regional) representatives. In Scotland as a whole 54 per cent of houses were owned by the state in 1978 ; the proportion is much higher in industrial areas, but recent government policy to allow local authority and New Town tenants to buy their houses may have very far-reaching effects on social patterns in Scotland.

Through the mechanism af Regional Reports and Structure Plans it is hoped that the new regional authorities will be able to produce a new type of plan which combines traditional British land-use planning with increased emphasis on economic and social factors. For the first time in Scotland there is a tier of local government responsible for the whole industrial area from Dumbarton to Motherwell (about 50 kms. direct). New integrated strategies have already been produced for public transport and education, although sometimes there is a suspicion that the new form of regional government is more popular with academic theorisers than with the actual people who are planned.

Although the Regional Council does not have direct responsibility for housing, it is inevitable that housing considerations have to be included as part of strategic planning. The Regional Council, then, has to consider whether it is in the regional interest to concentrate local and central government expenditure on areas of great need or on areas of relatively great potential (e.g. the New Towns). In general the present policy is to concentrate on the areas of need and this may be contrasted with the earlier central government policy of assisting growth areas in the mid-1960's.

The Labour government which came to power in 1974 made important changes in industrial and regional policy which have had direct effects on the Glasgow region. A major programme of dispersal of government offices from London was announced, although six years later relatively little has been achieved. Partly the delay may be related to a decline in the buoyancy of the London economy so that there is no longer a shortage of office accommodation in the capital, but trade union opposition to dispersal also seems to be very relevant. Glasgow has a very bad image in England (and perhaps in the world) as a violent and dirty city, and it has been difficult to persuade people from London to move north.

Much more important changes to industrial and regional policy were made in 1975 with the creation of the National Enterprise Board (NEB), the Scottish Development Agency (SDA) and, in 1976, the Welsh Development Agency. The N.E.B. was given wide powers to assist British industry but should probably be seen as part of the Labour government's industrial strategy, rather than its regional policy. The NEB, however, has intervened to assist key industries in distress in the West of Scotland (e.g. Rolls Royce and Chrysler) and the budget of the NEB has dwarfed those of the SDA and

W.D.A. The present role of the N.E.B. is not clear following the change to a Conservative government in 1979.

The Conservative Party, when in opposition, had argued that the creation of the S.D.A. would not be helpful and would lead to further nationalisation and bureaucracy, although in the parliamentary debates preceding the establishment of the S.D.A. Conservative spokesmen did not argue against provision of further government funds to assist industry in the Glasgow area. It seems unlikely that the S.D.A. will be disbanded in the near future, and as the main example of government intervention to assist industrial regeneration in the West of Scotland, the S.D.A. deserves close attention.

The main antecedent of the S.D.A. was probably the proposed SEDCOR recommended in the West Central Scotland Plan, but the creation of the S.D.A. should also be seen in the light of the apparent success of the Irish Development Authority (IDA) and the Highlands and Islands Development Board (HIDB). The I.D.A. is a national, rather than a regional, development agency, but its success in attracting foreign firms to Ireland through generous grants and tax incentives has received much attention. The H.I.D.B. had been set up by an earlier Labour government in 1965 in order to stimulate development in the North of Scotland, and has achieved success in improving economic conditions in some very remote areas. The H.I.D.B. operates mainly through a programme of selective assistance to industry (including fishing, agriculture, and tourism as well as manufacturing). A key feature of the I.D.A. and the H.I.D.B. has been that although government ministers retained overall control of policy and budget, the development institutions were allowed quite wide discretion in how budgets should be spent. In general both institutions reacted to proposals (by granting or rejecting requests for financial assistance) and have not indulged in the operation of business. The financial assistance has also been discretionary rather than statutory, i.e. each project had to be considered on its own merits.

The S.D.A. was given two main sets of powers relating to industrial support and environmental improvement. On the industrial side the S.D.A. was to act as an investment bank providing loan finance and equity—presumably to risky ventures which were not attractive to the private sector. The Agency was also to provide a financial advisory service to industry to undertake joint commercial ventures with private companies. Finally on the industrial side the Agency was to take over and continue the state programme of factory building in Scotland. In 1979-80 total expenditure of the S.D.A. was £93m., with receipts (e.g. factory rentals) at £17m.

On the environmental side the S.D.A. was given powers, and more important finance, to clear and improve derelict land. On the environmental side in particular, a good case could be made that the existing local authorities, particularly major Districts like the City of Glasgow, were perfectly competent to engage in such activity, but simply lacked the necessary funds. Nevertheless, all over Scotland, but especially in the Central Lowlands, industrial projects from the previous 200 years had left horrifying scars on the landscape, so that a major programme of environmental regeneration was certainly necessary.

Within Glasgow the most important part of the S.D.A.'s work has been its involvement in the Glasgow Eastern Area Renewal Project (GEARP). Much of the eastern part of central Glasgow had suffered on an extreme scale from the problems of the "inner city" that received prominence in England in the early 1970's. Much of the old economy had collapsed, widespread emigration had taken place, and many of the people who remained lived in wretched social conditions. The S.D.A. has been given a co-ordinating role in the GEAR project which aims to combine central and local authority funds in a comprehensive programme of new housing, environmental improvement, and industrial regeneration. Given the horrifying scale of the problem and the severity of general economic problems in Britain at present, the GEAR project seems to be achieving considerable success, for new factories are being tenanted, and decent housing is attracting people back to the east end of the city.

Glasgow in 1980

The population of Glasgow has declined drastically since the war, both absolutely and relative to Scotland and the U.K. The population was 1,090,000 in 1951 and is now about 794,000. In 1951 the Glasgow population was 21 per cent of the Scottish population and is now 15 per cent. Since 1972 Glasgow has experienced a net natural decrease in population.

Although many health indicators in Glasgow have improved significantly since the war, and infant mortality has declined dramatically, serious health problems remain. Both Scotland in general, and Glasgow in particular have very high death rates by world standards from heart disease and lung cancer. Air pollution is no longer a serious health hazard, and this is a tremendous improvement in less than a generation. The people of Glasgow do, however, persist in killing themselves slowly, through what is apparently the highest expenditure per household in Britain on both alcohol and tobacco.

Male unemployment in Glasgow has not been below 5 per cent since 1966 and has been above 10 per cent for the last five years. In November 1980 unemployment was 15.8 per cent for men and 9.1 per cent for women. Because of the inter-locking labour markets within the Glasgow region, unemployment rates are not available for sub-regions but clearly in some of the worst areas of Glasgow male unemployment must be at least 25 per cent. Fluctuations in Glasgow unemployment are closely related to fluctuations in unemployment in Scotland and Britain and it is little consolation that the disparity between unemployment levels in Glasgow and Britain seems to be declining. Clearly welfare measures introduced since the war have cushioned some of the impact of the dreadful unemployment rates, but the present level of unemployment seemed almost inconceivable in the 30 years after the war.

Accurate figures on income in Glasgow are not available, but it is probably relevant that Scotland is not a poor region by British standards. In 1977-78 the average weekly household income in Scotland of £100 per week was very slightly above the English level. This is not to deny that serious poverty exists in Glasgow, but there is an implication that low wagers in Glasgow are unlikely to be much of an attraction for new industry when Glasgow is compared with other locations in Britain. By Western European standards, however, Scotland is poor, with G.D.P. per head at about half the level of the Hamburg and Paris regions.[1]

In Western Europe car ownership (or the lack of it) is usually a good indication of income levels and in 1971 in Central

Clydeside (Glasgow plus ten surrounding Districts) 67 per cent of households did not have a car, while the corresponding figure for Great Britain is only 49 per cent (Fig. 11.1). It is perhaps surprising, then, that over the last 20 years the main improvements to the public transport system in Glasgow have been through motor-way construction. Such expenditure is usually justified on economic development grounds, although a programme of central government grants for motorway construction presumably influenced the priorities of local politicians. A re-built underground railway system was opened this year.

Housing in Glasgow has been bad by British standards for over 150 years. Huge improvements in the housing stock have been made since the war and present complaints sometimes seem almost trivial to people who knew Glasgow in the 1930's. Glasgow now has a surplus of housing (i.e. more house units than households) and this too is an almost unbelievable change from 25 years earlier. The emphasis in housing policy is now, therefore, far more on housing rehabilitation and general environmental improvement rather than simple construction of dwelling spaces. In addition the latest accurate statistics are for 1971, so that the exact scale of recent improvements and accurate up-to-date comparisons with other parts of Britain will not be possible until after the 1981 Census. In 1971, 16 per cent of Central Clydeside households lacked the basic amenities of exclusive use of hot water and/or fixed bath and/or inside water closet, and this is the same percentage as for Great Britain as a whole. Central Clydeside, however, did have very high levels of relative overcrowding—26 per cent of households at over 1 person to a room, compared to only 7 per cent in Great Britain as a whole. In 1971, 59 per cent of Central Clydeside households were in public sector rental (30 per cent in Great Britain as a whole) and almost all the post-war houses in Glasgow had been built by local or central government. During the last 10 years there has been intense ideological debate about the effects of such a policy.

Geographical class polarization has occurred with a movement to the suburbs of the middle-class comparable to the separation of black and white in American cities. It has been very difficult for tenants to make the transition from public sector to private sector housing. Now all the main political

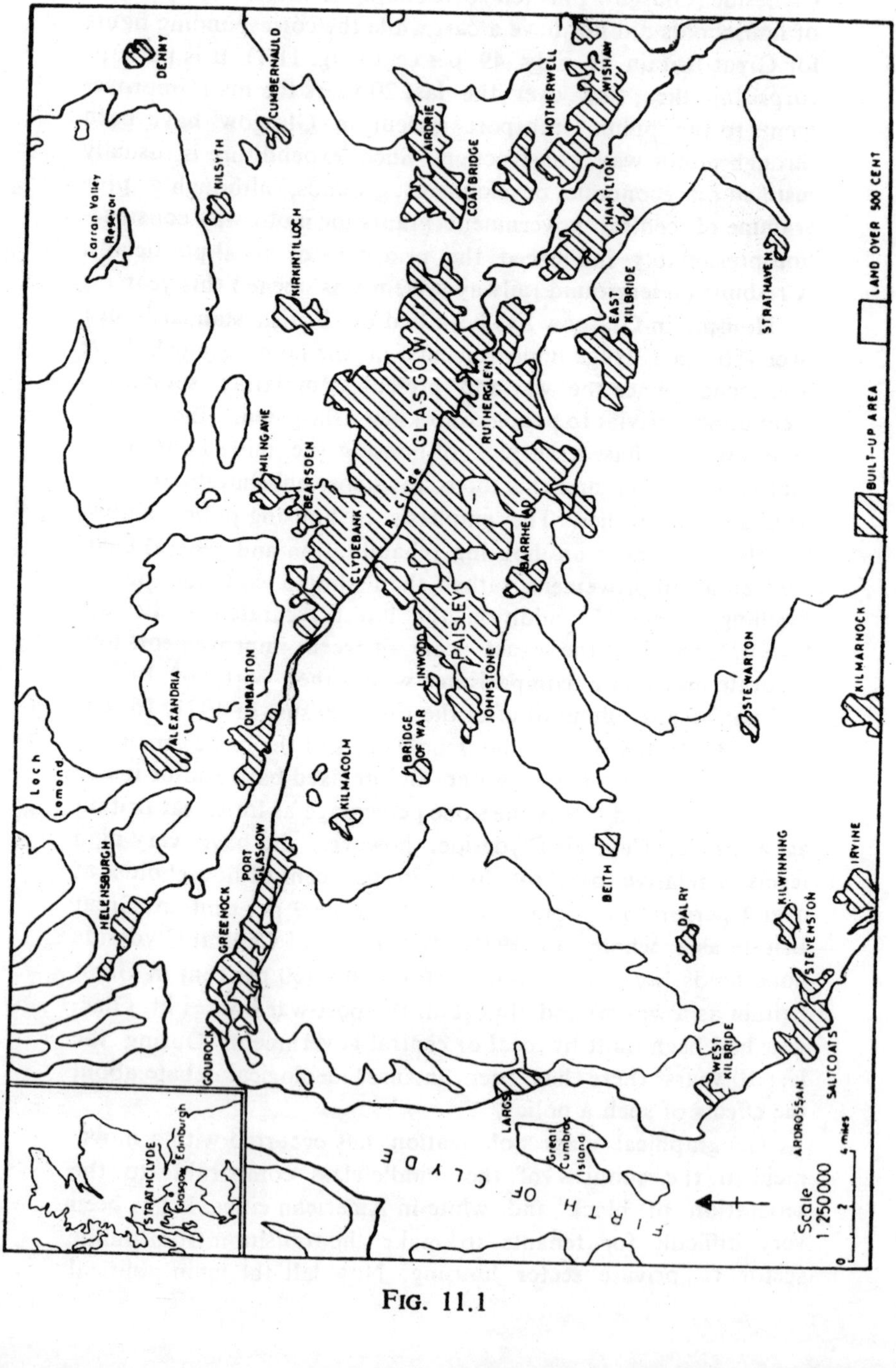
DENNY
CUMBERNAULD
KILSYTH
Carran Valley Resevoir
KIRKINTILLOCH
AIRDRIE
COATBRIDGE
MOTHERWELL
WISHAW
HAMILTON
EAST KILBRIDE
STRATHAVEN
GLASGOW
RUTHERGLEN
MILNGAVIE
BEARSDEN
CLYDEBANK
R. Clyde
BARRHEAD
PAISLEY
LINWOOD
JOHNSTONE
BRIDGE OF WAR
KILMACOLM
DUMBARTON
ALEXANDRIA
Loch Lomond
HELENSBURGH
PORT GLASGOW
GREENOCK
GOUROCK
BEITH
DALRY
KILWINNING
IRVINE
STEVENSTON
SALTCOATS
ARDROSSAN
WEST KILBRIDE
LARGS
Great Cumbrae Island
FIRTH OF CLYDE
STEWARTON
KILMARNOCK
BUILT-UP AREA
LAND OVER 500 CENT
Scale 1:250000
0 2 4 miles
STRATHCLYDE
Glasgow
Edinburgh

FIG. 11.1

parties in Glasgow are in favour of a policy of allowing local authority tenants to buy their houses, but the long term effects of the new policy are not clear.

Employment in the Glasgow region declined steadily over the period 1945 to 1970, but the process seems to have intensified in the last 10 years. Between 1951 and 1971 Glasgow experienced a net loss of some 90,000 manufacturing jobs and job losses were particularly severe in ship-building, metal manufacture, and textiles. Moreover, Glasgow had more severe job losses in these sectors than did the country as a whole, so a simple structural explanation is not sufficient. Now there are more jobs in services in Glasgow than in manufacturing. Particularly disturbing is the feature of a steadily increasing number of firms closing in the region (over the period 1950 to 1976), while the number of new firms opening has remained relatively constant.

CONCLUSIONS

One hundred years ago the living conditions for most people in Glasgow must have been atrocious, even if there had been improvements since earlier in the century. On the other hand the economic foundation of the region seemed secure. Severe economic depressions affected Glasgow in the nineteenth century, but continuous decline over a period of 30 years would have seemed most unlikely. Glasgow appeared to have real long-term advantages over the rest of the world in terms of engineering skills and entrepreneurial ability.

Now the living conditions for most people have improved almost beyond recognition, but there is great uncertainty about the economic future. In a real sense the people of Glasgow, through their suffering, made an essential contribution to the process of skill and capital accumulation that was part of the nineteenth century industrial revolution. Glasgow, however, somehow did not adapt to the change away from steam engines. The loss of markets in what had been the British Empire also seems to have been crucial, as perhaps was loss of local control of existing companies. It would be very difficult to make the case that the improvement in social conditions has in any way caused the deterioration in economic conditions.

Although the decline in Glasgow is not going to be reversed over-night, there are some optimistic features compared with only 10 years ago. The new system of Regional government seems to be a great improvement in terms of definition of objectives and allocation of resources according to what are perceived to be strategic issues. The S.D.A. has achieved much in the way of environmental improvement in only five years. Enormous housing improvements have been made since the war and the implications of an overall surplus of housing and a general increase in home ownership are not clear.

The key problems of the next ten years are clearly going to be concerned with economic regeneration. Probably the main justification for some optimism relates to the still great unexploited potential of North Sea oil. Apart from the at-present rather unlikely possibility of allocating oil revenues directly for regeneration of the Scottish economy, there is still the possibility that North Sea oil will offer great opportunities for industry in the Glasgow region. Fifteen years ago a forecast of huge increases in demand in Scotland for steel products in general, and naval engineering products in particular, would have seemed the salvation for Glasgow. This salvation has not occurred and is probably unlikely, but it is still possible.

FOOTNOTES

1. According to World Bank figures G.D.P. per head in the United Kingdom in 1978 was $5,030 compared with, to use the examples of the nations of conference participants, $2,380 in Yugoslàvia, $3,450 in Hungary, $3,670 in Poland, $3,700 in the U.S.S.R., $3,850 in Italy, $5,710 in the G.D.R., $7,030 in Austria, $9,090 in Belgium, and $9,580 in the G.F.R.

REFERENCES

Abercrombie, P. and Matthew, R. (1949) *The Clyde Valley Regional Plan*, Edinburgh, HMSO.

Checkland, S.G. (1977) *The Upas Tree : Glasgow 1875-1975*, Glasgow, University Press.

Cmnd. 2188 (1963) "Central Scotland : A Programme for Development and Growth", Edinburgh, HMSO.

Cunnison J. and Gilfillan J. (1958) eds., *The Third Statistical Account of Scotland : Glasgow*. Glasgow, Collins.

McCallum, J.D. (1979) "The Development of British Regional Policy" in Maclennan D. and Parr J.B. (eds.) *Regional Policy : Past Experience and New Directions*. London, Martin Robertson.

MaCallum, J.D. (1980) "Statistical trends of the British conurbations" in Cameron, G.C. (ed) *The Furure of the British Conurbations*, London, Longman.

Randall, J.N. (1980) "Central Clydeside—a case study of one conurbation" in Cameron, G.C. (ed) *The Future of the British Conurbations*, London, Longman.

Scottish Development Agency (1979) *Annual Report*, Glasgow.

Slaven, A. (1975) *The Development of the West of Scotland : 1750-1960*, London, Routledge and Kegan Paul.

West Central Scotland Plan (1974) *Supplementary Report 1, The Regional Economy*, Glasgow.

APPENDIX I

SOME GLASGOW STATISTICS

1. *Population 1801-1951*

Year	Area ('000 hectares)	Population
1801	2.0	77
1831	2.0	193
1851	2.0	329
1871	2.0	478
1891	2.4	566
1911	5.2	784
1931	11.8	1088
1951	15.9	1090

2. *Population 1951 and 1979 ('000)*

	Glasgow District	Strathclyde Region	Scotland	United Kingdom
1951	1,090	2,524	5,102	50,290
1979	794	2,431	5,167	55,883

3. *Perinatal Mortality Rates 1951 and 1977*
(No. of still births and deaths of infants in first week of life per 1,000 total births)

	Glasgow	Scotland	England & Wales
1951	48	44	38
1977	22	18	17

4. *Dependency Ratios 1951 and 1976*
(No. of people aged 14 and under plus people aged 65 and over as ratio of people aged 15-64)

	Glasgow	Scotland	United Kingdom
1951	1:1.99	1:1.88	1:2.01
1976	1:1.57	1:1.66	1:1.69

5. *Ratio of Glasgow unemployment percentage to Scottish and British unemployment percentage (July figures)*

	As % of Scotland		As % of Great Britain	
	Male	Female	Male	Female
1964	140	78	331	233
1969	129	80	207	178
1974	139	81	229	170
1979	123	96	168	144

6. *Manufacturing employment 1951-1976**

	Glasgow	Scotland	Great Britain
1951	223,000	804,200	8,382,100
1961	182,200	703,400	8,361,400
1971a	134,100	698,800	8,215 800
1971b	172,800	669,000	7,886,000
1976	144,200	608,000	7,246,000

*Figures up to 1971a are for the Glasgow City of County and use Census definitions of manufacturing employment ; figures from 1971b are according to Department of Employment definitions of Glasgow and manufacturing employment.

7. *Employment Structure 1976 (%)*

	Glasgow	Rest of Scotland
Primary	0.4	6.2
Manufacturing	30.4	25.0
Construction	7.1	8.5
Utilities & Transport	9.8	7.7
Distribution	13.4	11.3
Services	38.8	41.3

TWELVE

LEO R. ZRUDLO

THE ROLE OF CLIMATE AND CULTURE IN THE PLANNING OF SMALL SETTLEMENTS

THE planning of settlements has generally considered climatic and cultural factors in a superficial manner and in relatively general terms. This attitude has been even more flagrantly present when the planning of small settlements has been undertaken. In town or settlement planning some work has been done on the establishment of indoor comfort criteria but little or no concern has been given to outdoor comfort. This paper focuses on the climatic and cultural aspects of small settlements as it relates to outdoor spaces created by town or settlement layout.

Climate and planning

To ignore climate as a major influence on design is a common occurence as Lynch remarks : "It is unfortunately commonplace to import forms suitable for one climate into a completely different one : to plant lawns in the desert, to build North American houses in the tropics . . ."[1] and the author might add temperate-type houses in the Arctic. Cooper, discussing a government planning orientation for arctic settlements which accentuated heat conservation through the use of apartment buildings and row houses (ignoring cultural needs), felt that " . . . studies aimed towards arranging buildings in compact groups to take best advantage of winter sunlight and to afford the greatest protection in storms seems of more practical value."[2]

Given this attitude of using inappropriate models transferred from other climates and ignoring cultural needs, this paper attempts to outline a new approach to settlement planning. This approach proposes to integrate the requirements, specialists and users into the planning process in the spirit of collaboration through consultation and participation. The specialists have dominated the design process in the past, paying lip-service to climatic factors and almost completely ignoring cultural needs.

That town planning in the past has concerned itself with climatic factors can be witnessed by such historical examples as the town of Kahan in Egypt in 2,000 BC ; the principles of town layout in China during the Chou Dynasty ; the planning recommendations of Aristotle and Xenophon ;[3] the climatic considerations in planning proposed by Vitruvius[4] and in the 16th century laws set out for the location settlements in Latin America.[5] With the advent of the industrial revolution, there was a diminished concern with climate[6] because the process of urbanisation was more concerned with proximity to natural resources, energy sources and transportation routes. However as Wolf notes, " . . . the urban planner, since the origin of the profession in the late 19th century, traditionally operated from a set of visual or geometric presumptions about the proper, intelligent, beautiful, and/or hygienic arrangement of cities."[7] Planners were thus originally oriented towards the more formalistic concerns of geometry and syle. Wolf also observes that "The urban plan . . . began as a stylistic conceit derived from a number of selected prototypes which often boiled down to nostalgic cultural reminiscences and transmigrations of physical city-form patterns. The choice of geometric patterns for both the public spaces of the city and its principal buildings generally depended on the predisposition of the planner, the government, the era and the intended symbolism or message contained in the building or plan."[8]

Eventually planners became more concerned with other problems in city planning and re-directed the emphasis to the infrastructure. Wolf remarks that "Power and utility lines, drainage conduits, and the transportation network, all elements of the 'infrastructure' which has been ignored and taken for granted for so long, gained a position of prominence that far outweighed aesthetic considerations."[9] Stirling supports this opinion in a critique of town planning techniques, describing the planning process as follows : "A typical sequence of new town 'rational' planning is firstly the design of sewers/ services, 'part of the infrastructure', the routing of which dictates the layout of roads. This in turn, dictates the disposition of buildings and an abject environment results—particularly in relation to housing."[10]

More recently a concern for eliminating the negative effects of climate on man has led planners to adopt such planning solutions as enclosed shopping malls, gallerias, atriums, suspended pedestrian ways and underground shopping corridors. Although originally these planning devices were employed to protect inhabitants from noise and air pollution, they also eliminated the negative effects of excessive heat and cold, humidity and precipitation.[11] However, this protection has its price, which the energy crisis has dramatised, producing a group of architects, engineers and planners who are attempting to reduce the energy needs in contemporary buildings and cities. This has led to greater importance being placed on energy and climate as can be observed by the number of books being published on the subject for planners and architects.[13]

Building forms are frequently transferred from one climate to another which has few if any conditions in common,[14] and a poor microclimate often results, the effects of which "... people will suffer ... in silence, adapting to them as they can, sometimes even unaware of them."[15] However, a few communities have been planned where climatic factors were more appropriately emphasized. Halprin's study for Sea Ranch is an excellent example of the integration of climatic factors in the planning process[16] as are Erskine's plans for Svappavaara in Sweden,[17] Fermont in Quebec,[18] and Resolute Bay in the Canadian Arctic,[19] and to a lesser extent Fathy's plan for Gourna in Egypt.[20]

The town plans of Chandigarh in India, a few towns in New Jersey, Corcula in Dalmatia, Kitimat in Canada and

Medallin in Columbia all considered wind flow around buildings in their layout.[21] However, in these latter examples, the integration of climatic factors into the design process seemed to be subordinate to economic and social concerns, while the former examples (Halprin, Erskine and Fathy) do not use an explicit systematic method. Hellers observes that economy is the major criterion in the town planning design process, which results in a tendency to state climatic factors in economic terms as well.[22] Taesler noted that "... with the exception of air pollution control, very little if any serious attention is paid to local or urban climatic conditions..."[23] A good example of this attitude is McHarg's study of the metropolitan region of Philadelphia which identified air pollution, linked to temperature inversion, as the major climatic factor influencing planning strategy, with high summer heat and humidity as secondary climatic concerns.[24] His Staten Island study considers only the climatic data related to hurricanes and the resultant inundation as having any importance.[25] This minimal concern with climate has been due in part to the lack of systematic research into the town planning—climate relationship and any comprehensive climatic design method related to settlement design.[26]

Climate-planning design methods

Of the methods proposed for designing with climate, Olgyay's is the most well known. The method is however, more oriented to the design of buildings than town planning.[27] The manner in which the evaluated data is presented does not facilitate its use by the designer when planning community layouts. Olgyay's analysis for community planning[28] seems more readily carried out from data presented in tabular and textual form than from the elaborate graphical means he uses for house design. His examples of community plans are qualitative in nature and could have been duplicated by a sensitive designer without going through his complex and time-consuming procedure. The shelter design aspect of the plans are a little more quantitative, but with the exception of form, volume, and orientation considerations, could also have been achieved without the elaborate calculations required by the method. The notion of outdoor comfort is generally left rather vague and ambiguous, and his 'bioclimatic chart' for comfort[29] does not consider the effect of clothing, various rates of metabolic heat production and, with respect to cold climates like those of the nordic countries, does not concern itself with freezing and below freezing temperatures.

Givoni also formulated a design approach based on climatic factors, but it is mainly concerned with indoor comfort conditions and is therefore directed to building design rather than settlement planning.[30]

A design method developed by Koeningsberger, Mahoney and Evans is interesting in its systematic procedure which divides the design process into three stages : 1. sketch design stage ; 2. plan development stage ; 3. element design stage. In the sketch design stage the climatic data is analysed in 'Mahoney tables' giving stress indicators which when entered into a sketch design recommendations table indicate general layout suggestions such as "compact courtyard planning" etc.[31] The two other stages contain recommendations at the building and building element level and not at the settlement planning level. It would have been useful to have more quantitative recommendations at the level of settlement design with regard to height and spacing for solar shading or exposure, and air movement for natural ventilation. For settlement design, the notion of outdoor comfort should be related to the combination of physical activity, air movement, and solar radiation to be of any real value. The method is interesting because of its simple and straightforward procedure, but it is more related to the provision of indoor comfort conditions.

Ralph Knowles, in his Owen Valley study developed an ecological method which includes climatic factors in an integrative procedure. He defines what he calls 'stress domains' which are acted upon by forces such as ambient air temperature, incident energy, precipitation and mountain-valley lateral-thermal winds as well as geology, surface water, ground water and topography.[32] He assigns stress values to each domain according to the degree and number of forces acting upon it, which are termed the domain's stress coefficient. The built environment in the stress domain " . . . can be described as a function of the ratio between exposed surface and contained volume. The more the exposed surface for the contained volume, the more susceptible the arrangement. This surface to volume ratio (S/V) or the coefficient of susceptibility can be correlated with the stress range on the site and the susceptibility of a building can be matched to the stress in a domain."[33]

Knowles then establishes solar shading limits in relation to slope and orientation arriving at a height to area ratio (H/A) which establishes the maximum height of construction for a site without casting shadows on adjacent sites.[34] The land is then divided into increments according to the maximum dimensions allowed by the H/A ratio for the slope-orientation of the stress domain. Network analysis is used to establish values indicating the degree of optimization required for a site which sets out the acceptable range of S/V values. This is carried out for the slope-orientation construct to establish the range of H/A values acceptable for the site.[35]

At the organizational level, sites can be selected for more detailed study according to the S/V, H/A range, and the ratio of incident solar energy at the winter solstice to that of the summer solstice (E_w/E_s). The increment of development involving one or several buildings should, in combination with S/V and H/A values give an E_w/E_s ratio equal to 1 in order to justify building. He then demonstrates the process in the subdivision of a portion of the Owens Valley site by using the H/A ratio pyramid for the slope and orientation of the site at three levels of organization : order, structure and system.[36]

This method was developed as an ecological approach with the possibility of optimizing energy use, but this author feels that it is too complex to be of use to architects and planners at the scale of small settlement design. The amount of time that would be involved in the determination of the values in the network analysis system, in the optimization procedures before arriving at the very useful control tools (S/V, H/A and E_w/E_s) seems exaggerated for the degree of real accuracy that could be achieved.

Arens proposed a design method which integrates climatic factors and comfort notions in a systematic manner. He describes an eleven step procedure : 1. the identification of design parameters ; 2. independent variables 3. dependent variables ; 4. relationships among parameters and variables ; 5. prediction of independent variable values ; 6. identification of constraints governing dependent variables and design parameters ; 7. identification of values of design parameters 8. identification of expected values of dependent variables ; 9. investigation of the consistency of values ; 10. relationships and constraints ; 11. comparison of and selection from alternative sets of parameter values.[37] The method needs testing with a detailed case study and requires development for the design of comfort in outdoor spaces because Arens was more concerned with the relation of building performance to climate.

A method has been developed by the University of Minnesota's Underground Space Center based on McHarg's hand overlay transparency method. By rating climatic factors in varying degrees from best to worst (for example solar orientation is rated 1. south (best) 2. southeast 3. southwest 4. east-southeast 5. west-southwest 6. north, northwest, northeast (worst), factors such as solar orientation ; slope exposure ; albedo ; shadows cast by landforms ; northwest, southwest and south windsheds; other thermal effects ; etc. are mapped" with best zones in dark and worst in white tones. A composite map is made of all the factors, the darkest zone being the one best for building.[38] This method has merit, however, since it was developed for underground housing it has minimal impact on such factors as wind flow and snow accumulation in relation to groups due to a minimum of protruding elements which would normally modify the environment close to the ground. The method was developed to integrate what are called energy-related factors, many of which are climatic, but related to heat loss and solar heat gain, and are not concerned with outdoor comfort in public spaces and pedestrian ways.[39]

The various design methods referred to above, all have some drawbacks or disadvantages, therefore this author felt the need to develop a design procedure that would permit architects and planners to integrate climatic factors into the design process in a manner that would not require excessive amounts of time and complex data computations, but which would still retain a level of precision acceptable for preliminary planning needs.

Comfort

The notion of comfort can vary greatly from one climatic zone to another. In the Aictic, maximum solar and minimum wind exposure and snow accumulation control are desirable, whereas in hot-humid climates, minimum solar and maximum wind exposure are often necessary to creating comfort conditions. Therefore a generalisable planning model must consider both possibilities in climatic design i.e. protection from and augmentation of climatic effects.

Factors normally considered in thermal comfort are (1) the air dry bulb temperature ; (2) relative humidity ; (3) temperature of the skin ; (4) radiant temperature of surrounding surfaces ; (5) velocity of air movement ; (6) amount and type of clothing worn ; (7) amount of physical activity and the degree of acclimatisation to climatic conditions.[40] For factors 1 and 2 it is proposed to use the 'Mahoney tables' to organise and analyse the preliminary data which will give preliminary recommendations if the climate is hot-humid or hot-arid. If the climate is more temperate or nordic then the climatic data will be well organised for use with other analysis techniques. In regions with low temperatures the effects of wind and temperature dominate those of humidity and solar radiation.[41]

Factor 3—temperature of the skin, is eliminated because of the difficulty of correlating skin temperature with that of the ambient temperature. Steadman included values for skin temperatures in his calculations of clothing thickness related to comfort,[42] while Auliciems *et al.* did not use skin temperatures directly in their calculations of clo values.[43] Brown and Page point out that the skin temperature of the hand is higher for Inuit than Caucasians[44] indicating that there may be differences in skin temperature according to race or culture, further complicating the determination of general values.

Factor 6—amount and type of clothing—is of minimal importance because as Humphreys observes, man maximizes environmental comfort by "... choosing clothing appropriate to the microclimate (or) by choosing a microclimate appropriate to his clothing."[45] Without a statistical analysis of what constitutes average clothing for the average person for various seasons, it would be difficult to propose typical seasonal clo values. Clo values will be included in the Penwarden comfort curves in factors 4 and 5.

Factor 7—amount of physical activity and degree of acclimatization—will be considered only in terms of physical activity as they affect the Penwarden comfort curves and the windchill index in relation to distances travelled in a period of time under low temperatures and high winds. Acclimatization relates to the period of time an 'immigrant' spends in a new climate. Since acclimatization changes over time, it would be difficult to choose a stage of acclimatization and determine comfort conditions for that stage only.

Factor 4—radiant temperature of surrounding surfaces—will be considered in the preliminary stage in relation to surfaces that are in shade or full sun. In this sense the determination of zones of solar exposure and solar shading are important. Among the various methods of verifying the solar exposure/shading are mathematical formulae for solar altitude and azimuth,[46] solar charts,[47] heliodon[48] and computer shading programs.[49]

The mathematical formulae are precise, but the computations are time consuming when carried out for various buildings at various times of day and seasons, and accuracy depends on the precision of measurement and drawing instruments. The solar charts are less time consuming but have the same possible inaccuracies due to instrument and measurement error. The heliodon involves the construction of a model which is placed on a table, adjusted for hours of day, latitude, and season, upon which light is directed from a lamp simulating the sun. Inaccuracies are possible in the fabrication of the model and in the adjusting mechanism of the table. The computer program is the most accurate simulation device depending on the grid resolution possible on the display screen and/or graphic print-out plotter. The author proposes the use of a computer program such as Smith and Wilson's[50] which plots shadows and calculates shading percentages, or a heliodon if a computer program is not available.

Factor 5—velocity of air movement—is one of the most important in comfort considerations because the amount of heat carried away from the skin is directly proportional to the square root of the wind speed and the temperature gradient between the skin and the air.[51]

For temperatures at freezing or lower, the windchill nomogram should be used because it indicates the temperatures and windspeeds at which exposed flesh will freeze and the length of time to freeze. In spite of Wyon's objection to the wind chill index which considers only the cooling of exposed flesh and not of the whole body nor the effects of clothing or activity levels[52] the author feels that it should be used in preliminary design. (See Fig. 12.1.) The windchill index will provide a safety factor in the determination of comfort conditions because the factors not included will augment heat output or heat retention, and therefore the body will be comfortable at conditions worse than those indicated by the wind chill index.

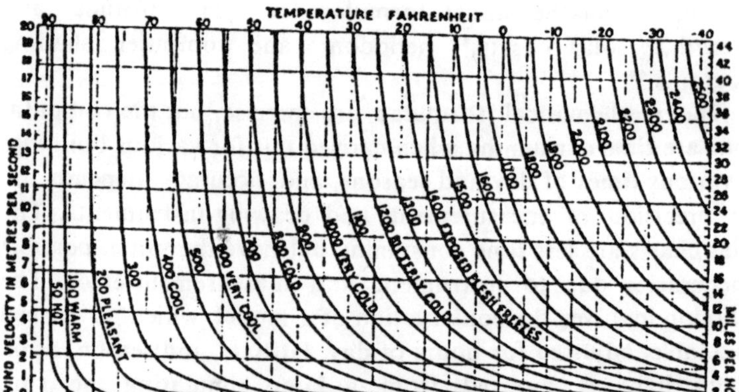

FIG. 12.1 : *Windchill Nomogram (from Baird, "Polar World", 1964)*

Penwarden's comfort curves are an interesting method for relating wind, temperature, solar radiation, clo and activity metabolic values to comfort, which would be useful for determining comfort in temperate and hot climates. The curves were produced using U.K. climatic data, clo values of 0 – 1.5 and a metabolic rate of 100 w/m^2 (strolling at 0.6 m/s 1.3 mph).[53] (See Fig. 12.2.) These comfort curves could be modified for other climates and for other activities such as sitting, standing,

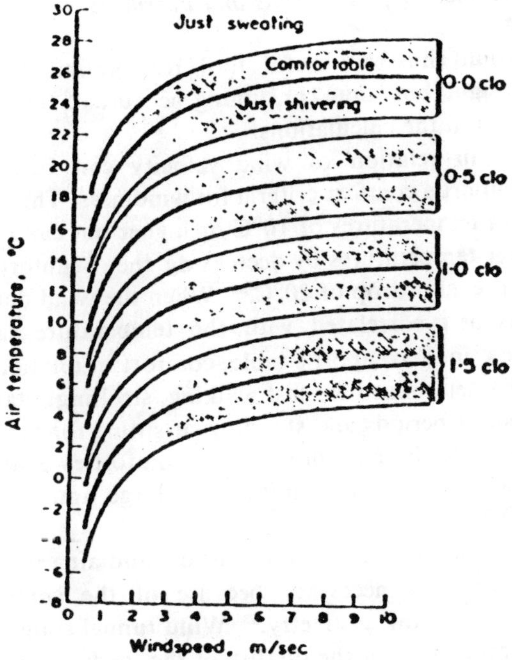

Comfort conditions for strolling in full sun.

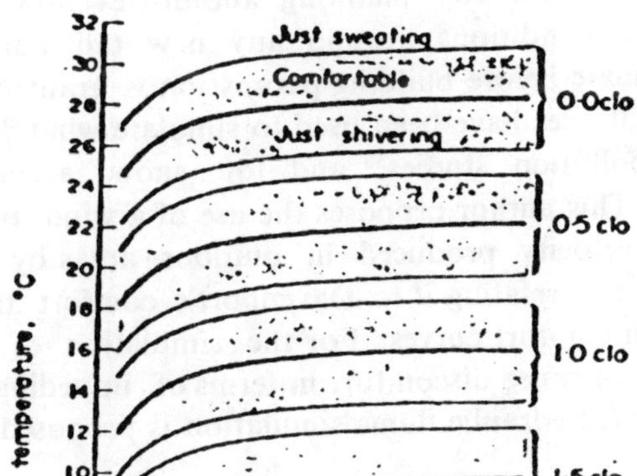

Comfort conditions for strolling in shade.

FIG. 12.2 : *Penwarden's Comfort Curves (From Penwarden, 1973)* etc., and would thus be a useful tool for preliminary exterior comfort design, but time consuming because of the amount of time involved in the calculations.

Comfort in relation to wind velocity can be determined using Davenport's comfort criteria for windiness. This data was produced for temperatures of 10°C or higher but can be adjusted for lower temperatures by decreasing the Beaufort number by one for every drop of 20°C.[54] When the wind data for the various seasons is correlated with the temperature range and wind speeds in Davenport's table, comfort in relation to wind speed can be determined for fast walking, strolling, standing or sitting for short periods and standing or sitting for long periods. This tool is useful because preliminary comfort evaluations can be made without the expenditure of large amounts of time. (See Fig. 12.3.)

For any accurate prediction of airflow and air velocity, wind tunnel simulation is necessary because of the complexity of urban environmental geometry.[55] Wind tunnel simulation as a tool in the prediction of the quality of the pedestrian environment created by proposed urban structures has gained importance.[56] Several city planning authorities now demand that the wind conditions around any new tall building be proven acceptable before building permission is granted.[57]

Hydraulic flumes have been used to simulate wind flow, but mainly for pollution studies[58] and for snow accumulation simulation.[59] This author proposes the use of a wind tunnel to verify wind velocity produced in outdoor areas by different urban layouts and relating it to Davenport's comfort tables or the Penwarden comfort curves. For the simulation of drifting snow, which can cause discomfort in terms of impeding pedestrian movement, hydraulic flume simulation is proposed.

Climatic factors—synthesis process

The specialists involved at this level—climatologists, architects, planners—after formulating the requirements and analysing the climatic data would generate plans and test them using the various simulation techniques described above. A specialist or group of specialists would be assigned the production of a plan for each climatic factor, i.e. solar, wind, snow.

Activity	Areas applicable	Relative comfort			
		Perceptible	Tolerable	Unpleasant	Dangerous
1. Walking fast	Sidewalks	5	6 (43%)	7 (50%)	8 (53%)
2. Strolling, skating	Parks, entrances, skating rinks	4	5 (23%)	6 (34%)	8 (53%)
3. Standing, sitting—short exposure	Parks, plaza areas	3	4 (6 %)	5 (15%)	8 (53%)
4. Standing, sitting—long exposure	Outdoor restaurants, band-stands, theatres	2	3 (0.1%)	4 (3 %)	8 (53%)
Representative criteria for acceptability			<1 occasion/week	<1 occasion/month	<1 occasion/year

Units : Beaufort number ; values in parentheses are frequencies with which wind speeds above 5 m/s occur, calculated as described in the text.

Temperature : >10°C ; at lower temperatures relative comfort level might be expected to be reduced by one Beaufort number for every 20°C reduction in temperature.

FIG. 12.3 : *Davenports Windiness Chart—Tentative Comfort Criteria* (*From Penwarden, 1973*)

The sun/shade or solar/shading aspects would be analysed and a plan made to achieve the degree of solar exposure or solar shading required for pedestrian paths, public gathering places and building facades for the various seasons. The plan would then be checked by a simulation technique (heliodon or computer solar shading program) and modified if necessary.

A second plan would be made based on the wind protection requirements determined by the Davenport comfort table or Penwarden comfort curves ; windchill nomogram ; and the Smith and Wilson wind reduction graphs for enclosed spaces.[60] (See Fig. 12.4.) Although these graphs were generated with scale models of walled enclosures, it is felt that the data will be useful for building groups that do not form a totally enclosed area, if the width to height ratio (W/H), (the width is measured perpendicular to the winds) of the enclosing building is reduced by one. This should minimize to a certain degree the difference between higher wind speeds at the rear corners of buildings that do not form walled enclosures and those that are walled enclosures. The length to height ratio (L/H), (the length is the distance between the front and rear buildings) is then used to determine the wind speed reduction percentage (V_{av}/V_{ref}) or the desired percentage wind reduction is chosen, and the length to height ratio is then determined.

Using these graphs, a first preliminary plan can be determined which would achieve the comfort levels proposed using the Penwarden comfort curves or the Davenport windiness table. The plan would be verified by simulation with a scale model of the layout in a wind tunnel. If the wind velocity measured on the pedestrian paths, public gathering places, playgrounds, etc. were not in the comfort range, then the plan would be modified and re-tested by simulation until the comfort levels were achieved.

If applicable, a snow accumulation control plan would be generated in a similar fashion using the downwind eddy dimension table produced by Evans,[61] which indicates the length of eddy in the lee of different building shapes. See Fig. 12.5. Snow accumulation on the lee side of buildings is due to the tendency of snow particles to deposit in zones where wind speeds are reduced to approximately 60% of the free-steam velccity.[62] For

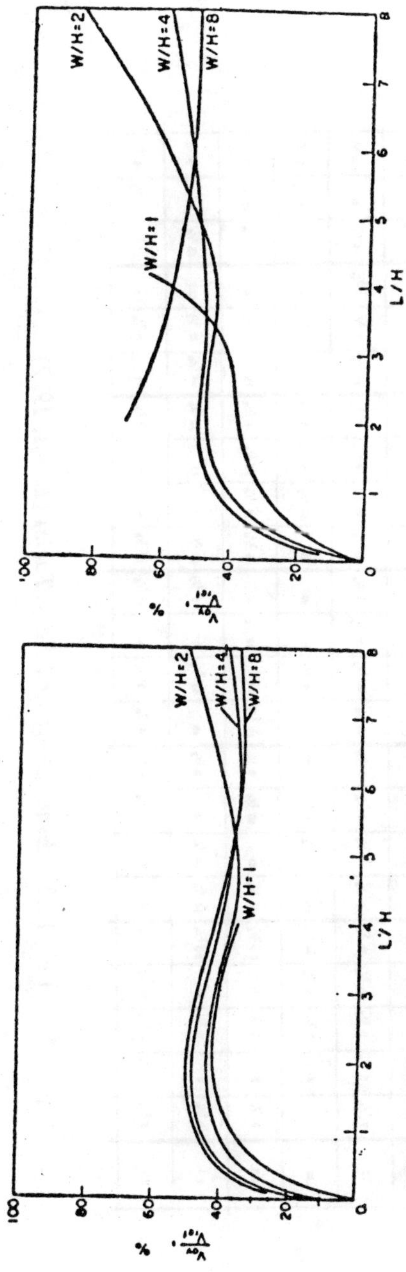

Effect of length on Average Wind Speeds at 0° orientation for Walled Enclosures

Effect of length on Average Wind Speeds at 30° orientation

FIG. 12.4 : *Smith and Wilson's Wind Reduction Graphs (Smith and Wilson, 1977)*

BUILDING TYPE AND ROOF PITCH

LENGTH (L)	W=1 H=1	2 1	3 1	W=1 H: 2	1 3	W=1 12/12	1 8/12	1 6/12	1 4/12	W=2 12/12	2 8/12	2 6/12	2 4/12	W=3 12/12	3 8/12	3 6/12	3 4/12	W=2 3/12	2 2/12	2 1/12	W=2 3/12	2 2/12	2 1/12
2	2½	2	2¼	8¼	6¾	3¼	3¼	2¾	2¾	2¾	2½	3	2½	3	2½	2½	2½	3	2½	2½	2½	2¼	2¾
4	3¾	2¾	3¼	8¼	11½	5	4¾	4¼	3¾	5¼	4¼	4	4¼	6	4¼	4½	3¾	5¼	4	3¾	4¼	4	4
8	5¼	3¾	4¼	11¾	16½	9	7¼	6¼	6½	9¼	8¾	8¾	6¼	10¾	8	6	6¼	8¼	7	6½	6½	6	5
12	7¼	5¾	5½	13¾	18¾	11¼	8¾	7¾	7¼	11¾	10¼	8½	7¾	10¼	10¼	9¾	8½	10¼	8¾	7¾	8¼	8	8¾
16	8	6	5¾	16¼	19¾	13¼	10¼	8¾	8¼	13¼	12½	10	10	15	12½	11¼	9¼	11½	10¼	10¼	11	9½	9½
20	8¼	6	8½	17¼	20¾	13¼	10	10¼	8¼	14¼	13	12¼	11	16¾	14¼	12½	11¾	13	11¼	11	11¾	10¼	8½
24	8¾	7	9½	18	20¾	14¼	11¼	11	9½	15	14¾	13	12½	17¾	15¼	13	12¼	14¼	12½	12	13¾	11	9
28	8¾	7¼	6¼	18½	21¼	16	12	12¾	9½	15¼	13¾	14	13¾	18½	16	15	12¾	15	13¼	13	13¾	11½	9½

FIG. 12.5 : *Evans Downuind Eddy Tables (Evans, 1972)*

a more detailed explanation of the procedure for snow control see Zrudlo.[63]

Once the plan has been drawn up to minimize the problems due to excessive snow drifting such as impeding vehicular and pedestrian movement between rows of houses, limiting the use of public gathering places and blocking entrances to buildings, a model of the layout would be constructed and tested in the hydraulic flume using a very fine sand to simulate snow particles.[64] The plan would then be modified or redesigned, depending on the simulation results.

The next step is the synthesis procedure begins by identifying the conflicting requirements in each plan in relation to the other two plans. Conflicts would be resolved by modifying the portion of the plan in conflict in such a way that it still satisfies all its own requirements as well as those of the other two plans. One of the plans may require less modification and could become the basis of a new composite plan satisfying the requirements of all the climatic factors. Where this is not possible a process of consultation, negotiation and compromise would be used to arrive at a plan that would satisfy the major or most important requirements. Simulation tests would be run on the synthesized plan to ensure that the major requirements of all the climatic factors are in effect satisfied. See Fig. 12.6 for a diagram of the synthesis procedure.

This iterative process could be shortened if the solar plan is formulated with the wind protection requirements in mind, which could possibly minimize the number of modifications to be made when the solar plan is tested for the wind protection requirements. If snow is a factor of concern, the plan based on snow accumulation control requirements would demand the most modifications because of the conflict created by snow accumulating in zones where wind speeds are the lowest.

The author believes that decomposition of the problem into its sub-components is essential, so that the analysis and the solution of the sub-components' problems can be kept separate so as not to influence immediately their individual solutions. If this is not done, there is the danger that an innovative plan for one of the climatic factors may be passed over or not emerge at all, because possible conflicts with the other climatic factors may be detected very early on. The conflicts should be eliminat-

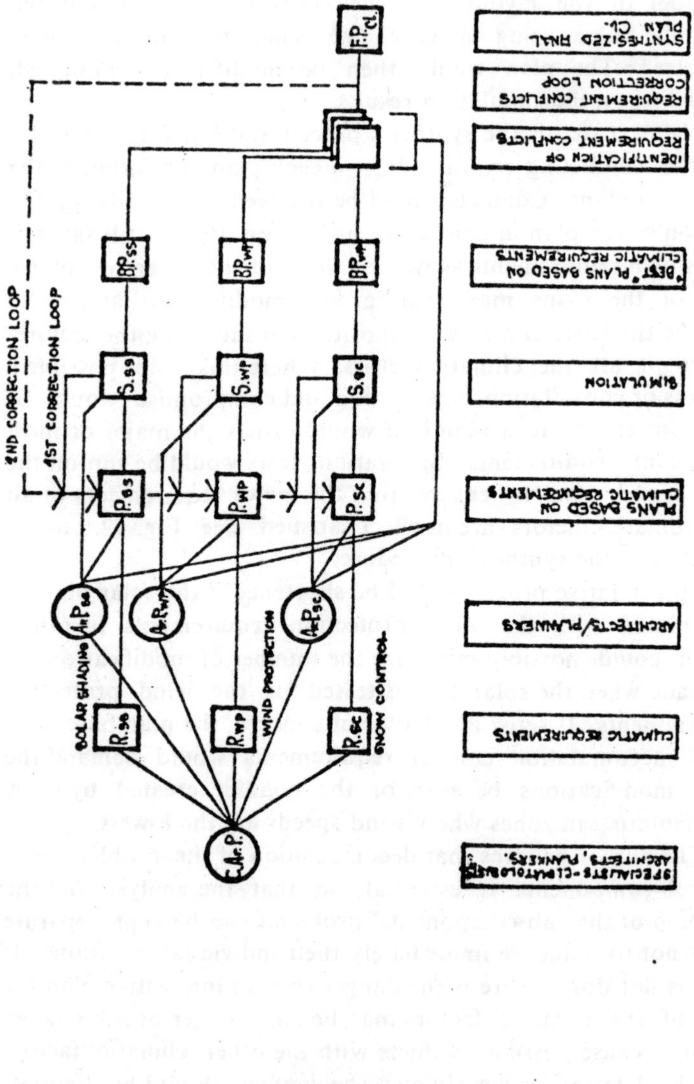

FIG. 12.6 : *Sgnthesis Process – Climatic Factors (cl.)*

ed at a later stage only after the 'best' plan for the particular factor is developed.

Ideally each plan is based on a separate climatic factor and should be developed by a different member of the design team concentrating on satisfying only the needs of his climatic factor, consciously eliminating any other influences. When the plans are compared, consultation and negotiation to produce a compromise or trade-offs must be made with the spirit of defending the requirements of each climatic factor. This is necessary to ensure as Alexander notes, the "... internal fitness..." in order "... to control its fit as a whole to the context outside."[65] Popper observes that "... many theories (or solutions) would be abandoned before their strengths were discovered if they were not defended strongly by their proponents."[66] However, in the author's model, once the final synthesized plan is agreed upon, all the team members would drop their partisanship for their particular plan and would support the synthesized plan.

The synthesized plan (Cl.) would then be synthesized in like manner with the synthesized plan of the geological/hydrological plan (G./A.) and the synthesis plan for the technical and social infrastructure plans (T.S.I./S.S.I.) which would be produced by specialists in the same manner as the climatic factors' plans, thereby completing the synthesis for the first three major factors.

Culture and planning

The cultural aspects of settlement planning have often received similar superficial and summary treatment as have climatic factors. Determination of desirable behaviour settings (i.e. "... small localities bound in time and space within which there is some stable pattern of purposeful human behavior"[67]) has received more attention than the more specific aspects of culture because most planning is involved with large cities which include several cultural groups. However, at the scale of small settlements, cultural groups tend to be more homogeneous and cultural factors necessarily should take on greater importance.

The dictionary definition of culture is "the way of life of a people", which more explicitly involves man's activities and beliefs. Rapoport defines the socio-cultural forces that should be reflected in living environments as "... religious beliefs, family and clan structure, social organization, way of gaining a livelihood, and social relations between individuals."[68]

Alexander defines two categories of cultural information in his example of the design of an Indian village : Religion and Caste, and Social Forces.[69] Lynch is more ambitious and proposes five different categories of cultural data for planning : 1. Resident and using population—which includes demographic, social, economic, political and forecast data ; 2. On-site behavior settings—socio-cultural data ; 3. Site values—legal, political and 'territorial' data ; 4. Past and future – historical and prospective data ; 5. Images—identification, meaning and preferences data.[70] Generally categories 1, 3 and 4 are well analysed by traditional planners. Category 2 has received much attention by environmental psychologists, but category 5 has been either ignored or treated superficially.

Architects and planners often show great interest in traditional societies[71] because meaning, identification and preferences can still be recognized in the spatial expression of their villages where their habitat is well integrated into the whole.[72] The social interaction in these societies is usually very simple. Alexander observes "In a traditional society, if we ask a man to name his best friends and then ask each of these in turn to name their best friends, they will all name each other so that they form a closed group. A village is made up of a number of separate closed groups of this kind.[73] Small settlements resemble traditional societies in that they are often made up of one cultural group, and as Lynch notes "Physical patterns may have predictable effects in a single culture, with its stable structure of institutions and values . . . Each culture has its own norms for city form, and they are independent of those of any other." It can be dangerous to develop generalisations about culture as they can often have hidden in them an "ethnocentric bias"[74] but as stated previously, small settlements are often composed of one dominant culture and as this paper has its focus on this type of settlement, the problem becomes simpler.

Culture—planning design methods

A variety of people have proposed methods for integrating cultural factors into the planning process. Steinitz has been studying existing urban environments to discover whether there are patterns of urban growth which present meanings that can define or explain culture.[75] Alexander proposes the use of patterns agreed upon and controlled by a hierarchy of groups involving the individual, family, clusters of families, neighbourhoods, communities and the towns, each being responsible for the patterns affecting their level of the town space.[76] Problems with the pattern language have been raised by several authors. Duffy and Torrey, for instance, bring out the fact that each pattern is culture-based and is therefore difficult to transfer from one culture to another,[77] and even though the patterns were further developed and are presented in "A Pattern Language"[78] as archetypes they still retain this problem as Ward points out, because it is difficult to prove that the patterns are in fact archetypal and therefore valid in any culture.[79] Protzen has more fundamental criticisms of which the three most important are : 1. the superficiality of the evidence supporting the patterns ; 2. the evidence presented by other studies, supporting certain patterns, are taken uncritically at face value ; 3. supporting evidence for a pattern is based on what Protzen calls a 'concensus theory of truth'.[80] He criticizes it mainly because it is an "all-encompassing theory" that cannot be refuted in part because each part or pattern is supported and cannot exist without other patterns upon which they depend, therefore the pattern language as a whole must be refuted.[81]

Settlement or town planning has become such a complex affair that only specialists seem capable of attacking the planning problems. Jean Nouvel observes that planners ". . . have surrounded their actions with a halo of hypercomplexity. They have drowned decisions : sociological analyses in geographical ones, geographical analyses in financial ones, and financial ones in scenarios for the future. Only super-specialists can understand so many things and so they alone can decide. They have thus acquired power which they guard jealously, and have become 'indispensable' advisors to the politicians."[82] Designers and clients are beginning to admit ". . . that expertise does not reside entirely in the designer but rather in all those whose interests are affected by a design problem"[83] and that ". . . by involving in the design process those who will be affected by its outcome, may provide a means for eliminating many potential problems at their source."[84]

User participation

The need to integrate users into the planning prccess is now supported by many architects and planners, as well as concerned citizens. The users of Les Halles in Paris state : "We believe that the monopoly of decision-making by an individual or a group of individuals is the negation of urbanity. We are in favour of broad consultation on the various points of view, especially those who will, day in and day out, suffer the consequences of the decisions taken."[85] Lynch observes that : "A highly decentralized decision process, in which the immediate users of a place make the decisions about its form, is a powerful ideal. It reinforces their sense of competence and seems more likely to result in a well-fitted environment, than if they were excluded."[86] The latter aspect is especially true when dealing with a different culture. Eastman noted that when dealing with users who have ". . . different cultural and behavioural patterns from those of the architect (or planner) . . ., the architect's own value system and intuition will lead him to false conclusions and result in the imposition of his values on the users."[87] The author has the same conviction as Alexander that ". . . the daily users of buildings (and towns) know more about their needs than anyone else" and just as in Alexanders's example of Ph.D. students in the "Oregon Experiment", there can be no doubt that one group of specific users knows more about the needs of another group of the same specific users than does a group of planners or designers.[88] Jones points out that " . . . there is a growing demand that all the people who are affected by a new design participate in the making of critical decisions."[89]

Fathy has a rather typical 'professional' attitude to participation which believes in limiting the scope of participation to the design of individual houses, but holds that the design of villages is the domain of the government and its specialists.[90] Nouvel proposes that "The myth of participation is nurtured though a posteriori public inquiries, once the studies have been carried out and the decisions taken. Such inquiries are based on dull, incomprehensible documents and ensure complete lack of public interest in architecture and town planning."[91] Cavdar suggests that the typical planning process by only informing the users of the outcome of the decisions taken by the planners and administrators, condemns them " . . . to be more objects of the planning process ; for the mosses to become the subject of the entire process an unobstructed and dialectical media of communication must be attained."[92]

The author believes that planners and designers should become " . . . technical advisors, providing a vast and subtle range of methods", as Nuttal suggests,[93] and as Lynch proposes, they should act as specialists ". . . in creating form possibilities, predicting their effects, and explaining how they can be technically accomplished, but his basic role is to disengage himself by communicating the necessary techniques of design and analysis and thus allowing the client to invent and build his own world." They would work ". . . to reveal hidden needs and possibilities . . ." and would ". . . promote and monitor specific environmental experiments so that new possibilities are opened up for the user, who can then, by his response to the experiment, classify his requirements both to himself and to others."[94] Pesce in discussing his Les Halles project for Paris concedes less control to the citizens, but nevertheless states the need for user input in planning projects : ". . . I'd like to bring in statements by the local residents, which may seem inflammatory, but I believe that a better product can result with their help than is possible through a single mind . . . it presumes that what the residents do will be watched by specialists who are not imposed from above, but who act as advisors."[95]

The planning approach the author proposes, would employ a form of user participation that would permit the formulation of town plans based on requirements as expressed by the future inhabitants. Therefore the cultural requirements, unlike those of the climate, would take the form of a plan generated by the users rather than proceeding by means of social and cultural analysis which would form the basis of requirements for the planner to meet. It is interesting to note a historical precedent for user participation at the scale of settlement planning in the form of the Oneidan community in the United States. This communistic society which existed between 1848 and 1880 involved the community in the design of buildings and the form of places as well as room decorations, etc.[96]

Scale model in user-designed settlements

Many authors have mentioned the problem of communication that exists between architects or planners and clients or users.[97] Broadbent suggests that scale models would help designers communicate with users more directly and more efficiently.[98] Scale models have beed used in various user-participation projects[99] and their use is widely accepted. The author proposes integrating the users into the design and decision-making process through the use of a scale model which would permit the inhabitants to design their own settlement.

Scale models for user participation in the design of settlements are not as common as their use in the design of houses. Houareau describes personality tests using village models[100] and a school designed with the Navajo people described in Arch. Forum[101] used a site model and a variety of building forms and modules that could be easily extended in use to the design of settlements. The Pluram planning group used a scale model to solicit settlement plans from the Inuit who were to be rehoused at a new site on Hudson's Bay.[102] The experience, unfortunately, did not integrate the users in the decision process as the plans produced were used as a guide to the final design produced by the planners. The number of people consulted was not necessarily representative of the whole community and their plans were not integrated into an overall design procedure where they would consult on all the plans produced, on the final plan representing their input, and the plan proposed by the planners.

The use of scale models would encourage and facilitate user participation in the design of the settlement. It is felt that the use of a scale model for soliciting information and integrating the user into the planning process is more accurate than most other methods that have been employed to discover users' needs. Interviews and questionnaires have many problems, as several authors have pointed out.[103] Broadbent notes that although many people believe that the verbal response given by a person expresses what he thinks, this is not always true.[104] Anthropologists working with the Inuit on the Nunaturliq project discovered that often the respondant was giving an answer he thought the interviewer wanted to hear.[105] Levin observed that "... the questionnaire,.. .tends to reflect the conceptual framework and attitudes of those who draw it up rather than those who fill it in",[105] which emphasizes the misconceptions and errors that can creep into a user participation project depending on the methods used.

SYNTHESIS OF CULTURAL PLANS

In the user participation process proposed by the author, a scale model of the village (composed of the site, all the houses and service buildings) would be used in consulting the population. The scale of the site and buildings should be sufficiently large to allow easy manipulation of the pieces, but still permit an entire family to gather around the model. The population would be divided into extended family or kin-related groups which should be small enough to consult easily on the design of the village. The plans generated in this way would be classified into typical plans and would be presented to the community at large where votes would be cast for the various alternatives. The plan receiving the most votes would be used in the final synthesis with the plans generated by the specialists. The user, therefore, would have an input into the generation of a typical plan, and a vote on the final selection—thereby being intimately involved in the decision-making process.

The kin-related group is an ideal consultation-design unit because a more natural consultation is possible amongst members of the same clan or kin group, and when differences of opinion arise this can be voiced without causing ill feelings. By consulting the kin-related or extended family group, the consulting group dominates in numbers the group of anthropologists, architects and/or planners conducting the scale model participation process. Also the familiarity within the group enables the responses to be less inhibited, providing the stimulus for open interaction. There is, however, the danger of a person with a strong personality dominating the consultation. This could deform the group's response, but by consulting all the family groups and making a synthesis of the plans into "typical" plans, such a deformation could be moderated in the synthesis process.

The extended family group would be sufficiently large to ensure a variety of opinions and ideas, but sufficiently coherent that a consensus could be reached within the group. It is important, however, to ensure that everyone in the group is consulted because quite often the women and young adults (18-29) are frequently left out or do not feel at ease in group discussions where the clan leader or family head act as spokesmen for the group. In this case, it may be necessary to form separate consultation-participation planning sessions with the women and the young adults. The opinions of the young adults are important to include as they will form the future decision-makers and in cultures in transition, they will be the ones whose image of their culture will change the most.

The various typical plans generated by the extended family groups and, where necessary, the women and yound adults, would then be presented and explained in detail to the community in a general meeting. The "typical" plans, by integrating the common aspects of all plans within a category, would remove personal identification from the plans and with it the attendant problems of partisan voting for plans produced by "strong" personalities. The voting members of the community (age 18 and over) would then be asked to vote by secret ballot for the plan of their choice. The plan receiving the most votes (the cultural plan Cu.) would then be used in the synthesis with the synthesized plan for the other factors (geological/hydrological (G./H.) ; the technical service and social service infrastructure (T.S.I./S.S.I.) ; and the climatic (Cl.). See Fig. 12.7 for the synthesis procedure for the cultural (Cu.) factors.

The author feels that a scale model would be more accurate in interpreting the needs of the users because the intervention of the specialists would be absent from this phase (except to classify the user-produced plans) of the synthesis process, thereby minimizing their influence on the population's response. The community would also become aware of individuals' priorities and criteria for a village layout by seeing plans produced by other members of the community, aiding them to accept plans not completely based on their requirements.

The village plan generated by the scale model would give the direction of the development of the settlement from the agreed upon starting point—the plan accepted by vote. The plan could, however, change as cultural changes develop and are expressed through the same form of consultation and participation at regular intervals such as every five to ten years. This is important for a culture in transition such as the indi-

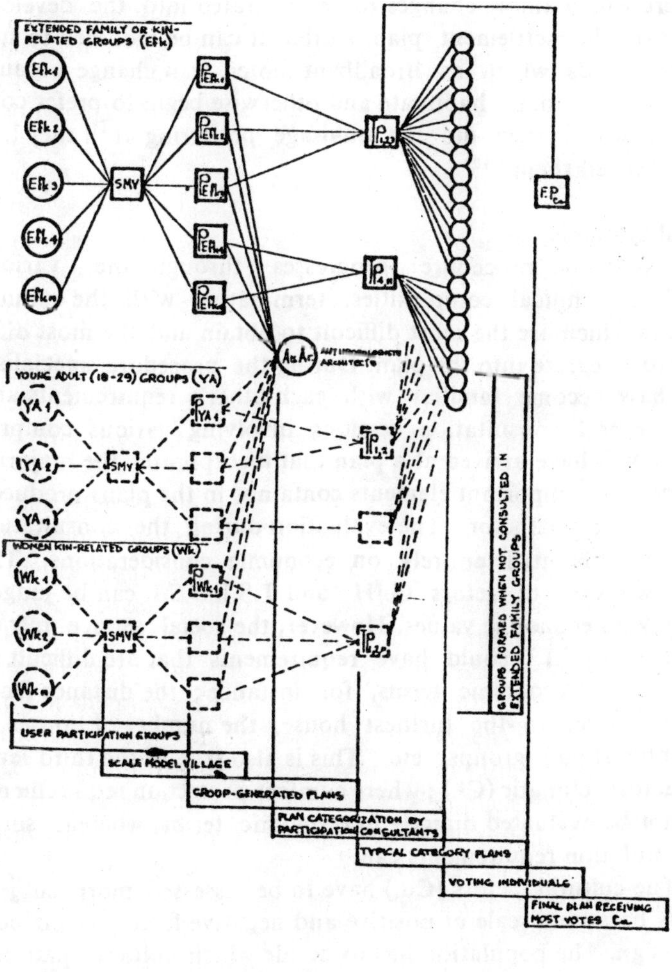

FIG. 12.7 : *Synthesis Process—Cultural Factors* (cu.)

genous peoples of the world. Fathy has pointed out the difficulty of designing for communities in the process of change[107] and Cavdar has observed changing patterns of use due to cultural evolution.[108] It is important, therefore, to provide a procedure which allows changes to be integrated into the development of the settlement plan so that it can be modified to suit future needs which as Broadbent notes, can change because people ". . . adapt, habituate and otherwise begin to prefer conditions which they did not envisage preferring until they had experienced them."[109]

Final synthesis

The synthesis procedure progresses through the various levels of technical complexities, terminating with the cultural factors which are the most difficult to obtain and the most difficult to integrate into the plan. During the procedure, specialists will have become familiar with each other's requirements and after several consultation sessions involving various compromises, will have arrived at a plan that incorporates the majority of the more important elements contained in the plans produced for the various factors. The evaluation during the consultation would not be made entirely on economic considerations. The first two levels of factors G./H. and T.S.I./S.S.I. can be judged mainly on economic values. However, the social service infrastructure (S.S.I.) could have requirements that are difficult to evaluate in economic terms, for instance; the distance from these services to the furthest house, the number of houses in neighbourhood groups, etc. This is also true of the third level of factors, climatic (Cl.) ; where comfort-protection requirements cannot be evaluated directly in economic terms, whereas snow accumulation requirements can.

The cultural factors (Cu.) have to be assessed more subjectively because a scale of positive and negative factors is difficult to assign. The population has to decide which cultural patterns they believe must be kept at all costs ; which ones can be modified slightly ; and which ones can be modified greatly or dropped. This graduation would permit dialogue and discussion which would eventually lead to agreement on an acceptable compromise in a plan retaining the more important factors, while modifying portions of the plan controlled by factors of lesser importance. It is also important not to bring the entire population into the consultation, negotiation, and synthesis of the first three levels, because such a repetitious procedure with so many people could lead to a lack of interest through exposure to too much information too many times. For this reason and because the cultural factors are the least technical in nature, they are integrated in the last phase of the synthesis process. A group designated by the community would, however, follow all phases of the synthesis procedure.

The final phase would involve the synthesis of the cultural factors plan with the synthesis of the plans at the three other levels :

(G./H. → T.S.I./S.S.I. → G./H.—T.S.I./S.S.I. → Cl. →
G./H.—T.S.I./S.S.I.—Cl. → Cu.)

Any conflicts resulting from the synthesis would be discussed with the specialists and elected representatives of the village who would defend the cultural plan in the negotiation process. After explanations by the community representatives and the specialists, concerning the advantages and disadvantages of the two plans (Cu. and the synthesis of the three other levels G./H.—T.S.I./S.S.I.—Cl.) in relation to the areas of conflict, negotiations would take place with compromise and trade-offs until a plan acceptable to the two groups was arrived at. The process of compromise could take some time if certain aspects of the cultural plan, which are in conflict with the synthesis plan of the other levels, are essential to cultural traditions. A trade-off might involve the acceptance by the community of higher installation or maintenance costs if a cultural pattern which is in conflict is deemed crucial.

When the final plan is ageeed upon, it would then be resubmitted to the people in the village for their comments on the compromises, if any were made. If a compromise was made that was not acceptable to the majority of the people, negotiations would be re-opened with the specialists to produce another plan which would resolve all objections. It would then proceed to the contract plan stage and to final implementation. See Fig. 12.8 for the organizational diagram illustrating the total synthesis prccess.

This planning approach, the author believes, will finally give the much neglected factors of climate and culture the impor-

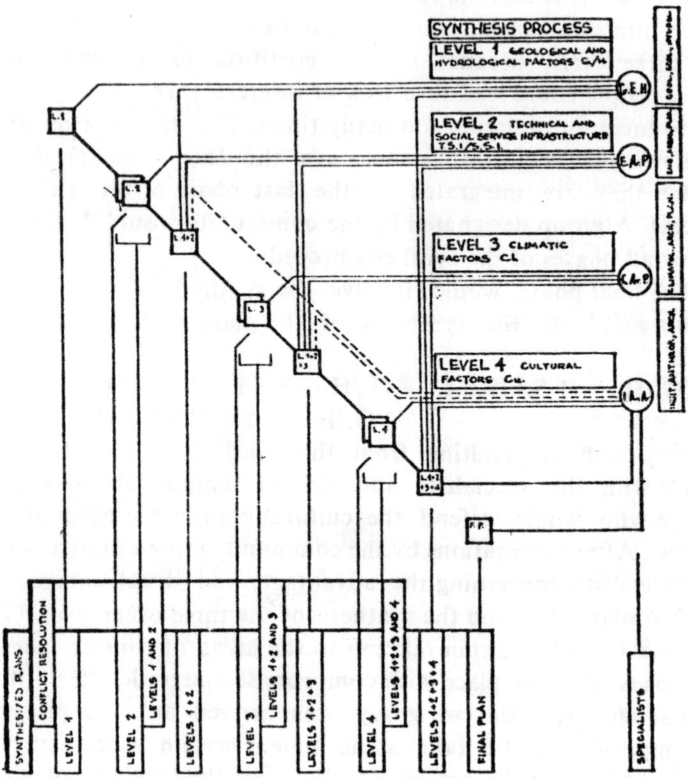

FIG. 12.8 : *Total Synthesis Process*

tance that they deserve in the planning of settlements. The approach encourages increased cooperation through the community members' better understanding of the "... assumptions, value-systems, and possible flaws of the system arrived at (and) they will be more willing to cooperate in carrying out the decision taken."[110] Thus the community would be more likely to accept compromises in the settlement plan when they have been involved directly in the decision process and understand the reasons for the various decisions. A village planned in this manner has a greater chance for success than those planned by the more traditional compartmentalized planning approach.

FOOTNOTES

1. Lynch, K. *Site Planning*, second edition, M.I.T. Press, Cambridge, Mass., 1971, pp. 16-17.
2. Cooper, P.F. "Applications of Modern Technology in an Arctic Environment", *The Polar Record*, Vol. 14, No. 89, 1968, p. 153.
3. Aynsley, R.M., Melbourne, W. and Vickerey, B.J. *Architectural Aerodynamics*, Applied Science Publishers, London, 1977. pp. 1-6.
4. Vitruvius *The Ten Books on Architecture*, translated by M.H. Morgan, Dover Publications, N.Y., 1960, pp. 25-26, 170-171.
5. Aynsley et al. 1977, pp. 6-7.
6. Ryd, H. "Use of Climatological Data in Building Planning with Respect to Comfort", *CIB/WMO Colloquium Teaching the Teachers Building Climatology*, CIB Proceedings No. 25, Document D20, the Nat'l Swedish Inst. For Bldg. Research, Stockholm, 1973, p. 55.
7. Wolf, P. *The Future of the City—New directions in urban planning*, Whitney Lib. of Design, N.Y., 1974, p. 163.
8. Ibid, p. 163.
9. Ibid, p. 119.
10. Stirling, J. "Nolli: Sector IV, Roma Interotta", *A.D.* Vol. 49, Nos. 3/4, 1979, p. 42.
11. Wolf, P., 1974, p. 122.
12. Ibid, p. 133.
13. The following is given as an example :
    Watson, D., *Energy Conservation through Building Design*, McGraw-Hill, N.Y., 1979. Univ. Minnesota, Underground Space Center, *Earth Sheltered Community Design*, Van Nostrand Reinhold, N.Y., 1981.
    Ridgeway, J. *Energy Efficient Community Planning*, J.G. Press, Emmaus, Pa., 1979.
14. Lynch, K. 1971, pp. 16-17.
15. Ibid, pp. 38-39 also.
    Dubos, R. "The Crisis of Man in his Environment", *Human Identity in the Urban Environment*, editors, G. Bell and J. Tyrwhitt, Penguin Books, Middlesex, England, 1972, p. 181.
16. P.A. "Ecological Architecture : Planning the Organic Environment", *Progressive Architecture*, May, 1969, pp. 120-137.
17. Erskine, R. "The challenge of the high latitudes", *R.A.I.C. Journal*, Toronto, January, 1964, pp. 33-41.
    Egelius, M. "Ralph Erskine : the humane architect", *A.D.*, Vol. 47, no. 11/12, 1977, pp. 93-95.
18. Van Ginkel Assoc., *Building in the North—Experience and Projects*, Vol. 2, Canadian Arctic Gas Study Ltd., March, 1976, pp. 8-9.
19. Culjat, B. *Climate and the Built Environment in the North*, Ph.D. thesis, KTH, Stockholm, 1975, pp. 226-294.
20. Fathy, H. *Architecture for the Poor—an experiment in rural Egypt*, Univ. of Chicago Press, 1973, pp. 70-71 and Fig. 66.

21. Aynsley et al. 1977, pp. 9-11 and Aronin, J.E. *Climate Architecture*, Reinhold, N.Y., 1953, pp. 211-221.
22. Hellers, B.G. 'Climate and Building in Sweden—Examples of ultimate design", *CIB/WMO Colloquium Teaching the Teachers Building Climatology Vol. 2*, Stockholm, 1972, pp. 31-1.
23. Taesler, R. "The Relevance of Urban Climate to Building and Planning", *CIB/WMO Colloquium Teaching the Teachers Building Climatology, Vol. 3*, Stockholm, 1972, pp. 69-1.
24. McHarg, I.L. *Design with Nature*, Doubleday and Co., Garden City, N.Y., 1969, p. 64.
25. Ibid, p. 105.
26. Givoni, B. "Effect of Town Planning on the Urban Climate", *CIB/WMO Colloquium Teaching the Teachers Building Climatology*, Vol. 1, Stockholm, 1972, pp. 18-2.
27. Olgyay, V. *Design with Climate—A Bioclimatic Approach to Architectural Regionalism*, Princeton Univ. Press, Princeton, N.J., 1963, pp. 12-13.
28. Ibid, pp. 153-177.
29. Ibid, pp. 17-26.
30. Givoni, B. *Man, Climate and Architecture*, Second edition, Applied Science-Publishers, London, 1976.
31. Koenigsberger, O., Mahoney, E. and Evans, M. *Climate and House Design—Design of Low-Cost Housing and Community Facilities*, Vol. 1, United Nations, N.Y., 1971, pp. 23-31.
32. Knowles, R.L. *Energy and Form, an Ecological Approach to Urban Growth*, MIT Press, Cambridge, Mass., 1974, pp. 49-65.
33. Ibid, p. 66.
34. Ibid, pp. 71-76, see also Knowles, R.L., *Sun, Rhythm, Form*, MIT Press, Cambridge, Mass., 1981, for a development of this principle for solar rights in community planning.
35. Ibid, pp. 77-80, 86-102.
36. Ibid, pp. 115-188.
37. Arens, E.A. *Climatic Factors in Planning and Environmental Design*, Ph.D. thesis, University of Edinburgh, 1972, pp. IV-84 to IV-88.
38. Univ. of Minnesota, Underground Space Center, *Earth Sheltered Community Design*, Van Nostrand Reinhold, N.Y., 1981, pp. 38-55.
39. Ibid, pp. 168-174.
40. Univ. of Michigan, *Environmental Evaluations, SER 2*, Univ. of Michigan, Ann Arbor, 1965, p. 96.
41. Steadman, R.G. Indices of Windchill of Clothed Persons", *Journal of Applied Meteorology*, Vol. 10, Feb. 1971, p. 674.
    Arens, E.A., 1972, pp. A11 1 : 1, 1 : 2.
42. Steadman, R.G., 1971, pp. 676, 679.
43. Auliciems, A., De Freitas, G. and Hare, F. *Winter Clothing Requirements for Canada*, Climatological Studies No. 22, Environment Canada, Atmospheric Environment, Toronto, 1973, pp. 8-9.

44. Brown, G. and Page, J. "The Effect of Chronic Exposure on Temperature and Blood Flow in the Hand", *Man in Adaptation—A Biosocial Background*, editor Y.A. Cohen, Aldine Publ., Chicago, 1968, pp. 175-176.
45. Humphreys, M.A. "Clothing and the Outdoor Microclimate in Summer", *Building and Environment*, Vol. 12, 1979, p. 142.
46. Lynch, K. 1971, pp. 69-73.
47. Bennett, R. *Sun Angles for Design*, Robert Bennett, Bala Cynwyd, Pa., 1978.
48. Koenigsberger, O., Ingersoll, T. et al. 1974, pp. 269-271.
49. Smith, F. and Wilson, C.B. "The Shading of the Ground by Buildings", *Building and Environment*, Vol. 11, 1976.
    Gero, J.S. *Computer Applications in Architecture*, Applied Science Publishers, London, 1977, pp. 285-303.
    Rogers, R., Atherton, P., Nall, D., and Greenberg, D. "A Means for Including Shadowing in a Building's Thermal Analysis", *CAD 78*, IPC Science and Technology Press, 1978, pp. 97-109.
50. Smith, F. and Wilson, C.B., 1976.
51. Sargent, F. and Tromp, S. *A Survey of Human Biometeorology*, WMO Technical Note No. 65, WMO—No. 160, TP 78, Secretariat WMO, Geneva, 1964, p. 28.
52. Wyon, D.P. "Environmental Comfort—An Appraisal of Current Criteria and Future Trends", *Integrated Environment in Building Design—Proceedings*, Conference, Univ. of Nottingham, editor Sherrat, 1974, p. 24.
53. Penwarden, A.D. "Acceptable Wind Speeds in Towns", *Building Science*, Vol. 8, 1973, pp. 263-266.
54. Davenport, R.G. "An Approach to Human Comfort Criteria for Environmental Wind Conditions", *CIB/WMO Colloquium, Teaching the Teachers Building Climatology*, Vol. 1, Stockholm, 1972, pp. 8-4 to 8-6.
55. Aynsley, R.M. "Effects of Airflow on Human Comfort", *Building Science*, Vol. 9, 1974, p. 93.
56. Lemberg, R. *Wind Tunnel Study of Vancouver Square Project, Vancouver, British Columbia—Part 11 Detailed Report*, LTR-LA-125, Nat'l Aeronautical Establishment, N.R.C., Ottawa, 1973.
57. Hunt, J.C.R., Poulton, E.C. and Mumford, J.C. "The Effect of Wind on People : New Criteria based on Wind Tunnel Experiments", *Building and Environment*, Vol. 11, 1976, p. 27.
58. Vadot, L. *La Pollution Atmospherique, ses mecanismes et son Etude en Laboratoire au moyen de Modeles Hydrauliques*, Direction Scientifique d' Alsthom, Techniques des Fluides, Grenoble, 1977.
59. Krasinski de, J.S. and Anson, W.A. *A Study of Show Drifts Around the Canada Building in Calgary*, Dept. of Eng., Univ. of Alberta, Calgary, Report No. 71, 1975.
    Calkins, D.J. *Simulated Snowdrift Patterns—Evaluation of Geometric Modeling Criteria for a Three Dimensional Structure*, Corps of Eng.,

U.S. Army, Cold Regions Research and Engineering Laboratory, Special Report 219, Hanover, N.H., 1975.

Frenette, M. "Simulation de Tempete de Neige en Laboratoire" No. 330, mars/avril, 1979.

60. Smith, F. and Wilson, C.B. "A Parametric Study of Airflow within Rectangular Walled Enclosures, "*Building and Environment*", Vol. 12, 1977, pp. 223-230.

61. Evans, B.H. "Natural Air Flow Around Buildings", *CIB/WMO Colloquium, Teaching the Teachers Building Climatology*, Vol. 1, Stockholm, 1972, pp. 12-1 to 12-13.

62. Melbourne, W.H. and Styles, D.F. "Wind Tunnel Tests on a Theory to Control Antarctic Drift Accumulation Around Buildings", reprint *of Proceedings of Int'l Research Seminar on Wind Effects on Bldgs. and Structures*, Ottawa, 1967, pp. 144, 153.

63. Zrudlo, L.R. *An Integrated Design Approach to Settlement Planning in the Arctic*, Ph.D. thesis, Univ. of Edinburgh, 1982, pp. 227-237, 282-290.

64. Calkins, D.J. 1975, and Krasinski and Anson, 1975.

65. Alexander, C. *Notes on the Synthesis of Form*, Harvard Univ. Press, Cambridge, Mass., 1964, p. 18.

66. Popper, K.R. *Conjectures and Refutations, the Growth of Scientific Knowledge*, Routledge and Kegan, London, 1963, p. 312.

67. Lynch, K. 1971, p. 13.

68. Rapoport, A. *House, Form and Culture*, Prentice Hall, Englewood Cliffs, N.J., 1969, p. 47.

69. Alexander, C. 1964, pp. 136-138.

70. Lynch, K. 1971, pp. 93-94.

71. Books such as :

Oliver, P. *Shelter, Sign and Symbol*, Overlook Press, Woodstock, N.Y. 1977.

Duly, C. *The Houses of Mankind*, Thames and Hudson, London, 1979.

Denyer, S. *African Traditional Architecture*, Heinemann, London, 1978.

72. Allen, E. *Stone Shelters*, MIT Press, Cambridge, Mass, 1969.

73. Alexander, C. "A City is not a Tree", *Human Identity in the Urban Environment*, Penguin Books, Middlesex, England, 1972, p. 412.

74. Lynch, K. *A Theory of Good City Form*, MIT Press, Cambridge, Mass, 1981, p. 101.

75. Wolf, P. 1974, p. 191.

76. Alexander, C. *The Timeless Way of Building*, Oxford Univ. Press, N.Y. 1979, pp. 505, 506.

77. Duffy, F. and Torrey, J. "A Progress Report on the Pattern Language", *Emerging Methods in Environmental Design and Planning*, edited by G.T. Moore, MIT Press, Cambridge, Mass, 1970, p. 264.

78. Alexander, C., Ishikawa, S., Silverstein, M., Jacobsen, M., Fiksdahl-

King I., and Angel, S. *A Pattern Language*, Oxford Univ. Press, N.Y., 1977.

79. Ward, T. "Tony Ward Reviews : A Pattern Language", *A.D.* Vol. 49, No. 1, 1979, p. 17.
80. Protzen, J.-P. "The Poverty of the Pattern Language", *Design Methods and Theories*, Vol. 12, no. 3/4, Sept.-Dec., 1978, pp. 193, 194.
    Rabeneck, A. "Andrew Rabeneck Reviews : A Pattern Language", *A.D.* Vol. 49, No. 1, 1979, p. 20.
81. Protzen, J.-P. 1978, p. 194.
82. Nouvel, J. "Impossible Urbanity", *A.D.* No. 11/12, 1980, p. 14.
83. Sanoff, H. *Methods of Architectural Programming*, Hutchinson and Ross, Stroudsberg, Pa., 1977, p. 167.
84. Cross, N. "Here Comes Everyman", *Design Participation*, Academy Editions, London, 1972, p. 6.
85. Miotto, L. and Muret, J.—P. "Urbanity in History" *A.D.* No. 11/12, 1980, p. 13
86. Lynch, K., 1981, p. 44.
87. Eastman, C.M. "Adaptive—conditional Architecture", *Design Participation*, London, 1972, p. 52.
88. Alexander, C., Silverstein, M., Angel, S., Ishikawa, S., and Abrams, D. *The Oregon Experiment*, Oxford Univ. Press, N.Y., 1977, pp. 40, 48.
89. Jones, J.C. *Design Methods—Seeds of Human Futures*, Wiley-Interscience, London, 1970, p. 71.
90. Fathy, H. 1973, p. 33.
91. Nouvel, J. 1980, p. 14.
92. Cavdar, T. "Design Participation as a Tool Towards Mass-Consciousness—Izmit Innovative Settlements Project", *Design Methods and Theories*, Vol. 13, No. 3/4, 1979, p. 163.
93. Nuttal, J. "How to Use Technology", *Design Participation*, Academy Editions, London, 1972, p. 20.
94. Lynch, K. 1971, pp. 259, 260.
95. Pesce, G. "A Certain Idea of France", *A.D.* No. 9/10, 1980, p. 24.
96. Lynch, K. 198, p. 68.
97. Dluhosch, E. "What if.... Some Thoughts on Methodology, Technology and the Scientific Approach ... Viewed Through the Looking Glass of Berkeley", *DMG-DRS Journal Design Reseach and Methods*, Vol. 7, No. 3, 1973, p. 223.
    Bayazit, N. "Comparison of Users' and Architects' Behaviours in Subsidized Co-operative Housing in Turkey", *Design Methods and Theories*, Vol. 14, No. 1, 1980, p. 17.
    Broadbent, G. 1973, p. 220.
98. Broadbent, G. 1973, p. 221.
99. Spring, B.P. *Planning and Design Workbook for Community Participation*, Princeton Univ., Princeton, N.J., 1969, pp. 404, 591.
    P.A. "3-D Scale Models" *Progressive Arch.* Vol. LV, No. 1, 1974, p. 96.

Saladin d' Anglure, B., Larochelle, G., Zrudlo, L.R., Baillargeon, R., Poulin, O., Berry, J., Qumaluk, N., Mark, J. and Tribouled, E. *Nunaturliq Phase-1, 1973*, Univ. Laval, Quebec, 1974.

Schmidt, J.J. "Participatory Design with a Scale Model—A Primer", *Conflicting Experience of Space*, Proceedings of the 4th Int'l Architectural Psychology Conferences, Catholic Univ. of Louvain, Louvain-la Neuve, Belgium, 1980.

100. Houareau, M.-J. "Le Test du Village—Un Jeu de Construction qui eu dit long sur notre personalite", *Psychologie*, No. 44, 1973.
101. Arch. Forum, "Navajo School : A Study in Community Control, *Arch. Forum*, Vol. 137, No. 2, Sept. 1972, pp. 54-57.
102. Pluram Inc. *Analysis of the Inuit's Relocation from Great Whale River to Richmond Gulf*, Makivik Corp., Pluram Inc., Montreal, Dec. 1979, pp. 122-137, 484-522.
103. Siddal, J. "Value Theory as a Vehicle for User Participation in Design", *Design Participation*, Academy Editions, London, 1972, p. 95.

     Dluhosch, E. 1973, p. 223.

     Jones, J.C. 1970, pp. 222-224.
104. Broadbent, G. 1973, p. 120.
105. Saladin d' Anglure, B. et al. 1974.
106. Levin, P. "Participation Planning Decisions", *Design Participation*, Academy Editions, London, 1972, p. 35.
107. Fathy, H., 1973, p. 53.
108. Cavdar, T., 1979, p. 162.
109. Broadbent, G., 1973, p. 122.
110. Kreimer, A., Polydorides, N. and Wormhoudt, D.T. "A Theory of Critical Reflections in the Planning Process", *Design Methods and Theories*, No. 3/4, 1978, pp. 166-168.

THIRTEEN

GERALD H. KRAUSSE

JABOTABEK
Planning for Jakarta 2000

DURNIG the next two decades the first supermetropolitan city of Southeat Asia will emerge in western Java, joining the world's largest conurbations with a minimum population of 10 million inhabitants. This polynuclear system will encompass the city of Jakarta, the surrounding satellite towns of Bogor, Tangerang, Bekasi, Depok and many smaller semi-urban settlements (Figure 13.1). This conglomeration of urban centers, dubbed JABOTABEK (Jakarta, Bogar, Tangerang, and Bekasi) is perceived in response to a series of problems facing the continued expansion of Metro-Jakarta (Daerah Kusus Ibukota or DKI). For example, in 1978 the population of JABOTABEK was estimated to have been 10.4 million ; by the year 2000 it is expected to have more than doubled to between 20-24 million. For the period 1963-1970 the maldistribution of income increased substantially between Jakarta and the rest of Java.[1] Studies indicate that 16 per cent of Jakarta's population is classified as very poor, and 50 per cent as poor ; the equivalent figure for Botabek are approximately 40 per cent and 75 per cent respectively.[2] Within DKI Jakarta and the adjacent provinces of Botabek about 30,000 hectares of agricultural land was lost to development in the 1970's, less than 5,000 hectares of this land has been developed according to specific plans.[3] Probably the worst problem confronting the regional government is that of housing. Some 80 per cent of the city's housing stock is either of semi-permanent or

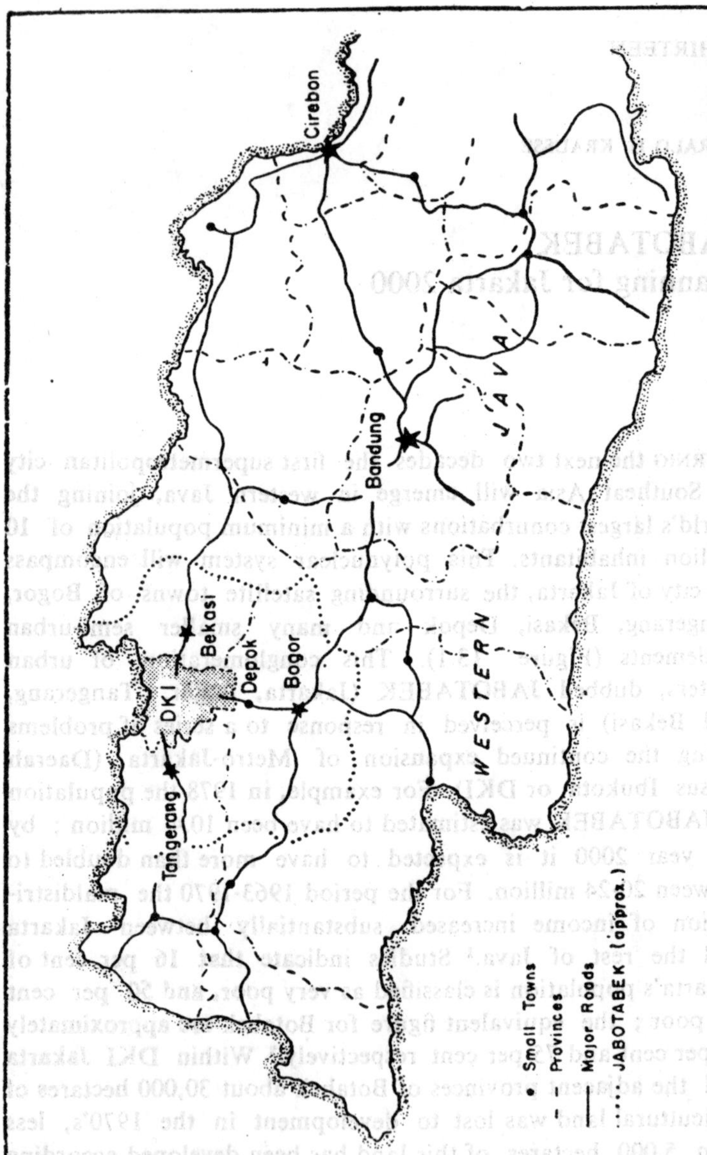
Cirebon
Bandung
Bekasi
Depok
Bogor
DKI
Tangerang
WESTERN JAVA
Small Towns
Provinces
Major Roads
JABOTABEK (approx.)

FIG. 13.1

temporary building material and is most frequently located on land whose legal ownership status in unclear. There is also a wide spectrum of issues affecting the supply and quality of services to the expanding population. In the case of drinking water, 90 per cent of the region's population obtains water supplies, untreated, from rivers, canals and wells ; tkose receiving treated water are almost all located in DKI. Only 24 per cent of the Jakarta residents have an electricity connection, compared to much higher rates in urban Java and a nation-wide urban average of 31 per cent.[4] Until recently government authorities have been unprepared to address these and related issues to spatial disparities and resource allocation as they pertain to the future development of JABOTABEK, particularly Botabek. Yet, the Primate City of Indonesia has undergone rapid economic growth and physical change which calls for major public and private action, the JABOTABEK Project is the most comprehensive planning document to date designed to deal with the following development objectives.[5]

1. Reduce the stress on the Capital City caused by rural migration and industrial as well as trade activities.
2. The development of fringe settlements (Bogar, Bekasi, etc.) as growth centers and residential locations.
3. The provision of infrastructure and land use control so as to stimulate investment in these growth centers.
4. Improving service delivery systems and income disparities for all population groups in the Jabotabek region.
5. Determination of resources to be committed during Pelita III and IV for all jurisdictions and levels of government.
6. The formulation of alternative development strategies and spatial plans according to the resources available.

Origin of the planning process

In 1973 it was realized that the physical expansion of Metropolitan Jakarta will eventually encompass most of the neighboring kabupaten. As a possible solution a "deconcentration strategy" was adopted, hoping to relieve pressures on the city. The Ministry of Internal Affairs, in 1974 submitted a report to President Suharto outlining the expansion of Jakarta based on the JABOTABEK concept. Upon presidential approval a team of experts (the Jabotabek Coordinating Board) was established to suggest appropriate organizational structures and projects necessary to implement the JABOTABEK Plan. The plan is currently carried out by the Directorate General of Housing, Building and Planning (Cipta Karya), the Directorate of Urban and Regional Planning (Tata Kota dan Tata Daerah) ; and is assisted by the Ministry of State (BAPPENAS), the Ministry of Public Works (Pekerjaan Umum) and the Ministry of International Affairs (Dalam Negeri).

Between 1973 and the present time the JABOTABEK Plan has experienced several conceptual and spatial changes. The 1973 plan, for example, focused on all existing centers surrounding Jakarta ; the 1975 proposal was scaled down to concentrate only on the development of three major towns in the fringe areas of the city. Even today, no attempt has been made to delineate the JABOTABEK region. There are many administrative divisions related to JABOTABEK depending upon what sector or type of service is involved. Data have been collected for DKI Jakarta, kotamadya Bogar and the kabupaten of Bogar, Bekasi and Tangerang. However, this area is larger than the expected expansion of JABOTABEK and should not be interpreted as the administrative region for the metropolitan organization. While the kabupaten level of administration is most useful for statistical purposes, the actual delineation of JABOTABEK will only include the existing centers of Serpong, Tangerang, Bakasi, and Depok.

In addition to the spatial plan concepts (discussed later), there have been many sectoral studies over the years. They include studies on water resources ; major roads and highways ; land use and traffic flow ; drainage, sanitation and flood control ; agriculture ; housing and services. The scale, socioeconomic starting point and approach vary substantially among the projects. Furthermore, the actual development trends differ greatly from the assumptions made in many sectoral studies. While these studies were made some years ago JABOTABEK has been growing at an ever increasing rate to approximately 280,000 people per year at the present.

A conceptual debate

In Indonesia the conceptual interpretation of the city's role in economic development divides urban planners into the "modernizers" and the "traditionalists". The modernizers argue that continued concentration of economic growth in the largest city is necessary to achieve economies of scale and infrastructure which in turn are the prerequisites for further growth needed to overcome societal disparities. Jakarta, they say, was the center of colonial penetration and today is a city of the national heartland which enjoys the existence of overhead facilities, external economics, political power and a variety of spatial advantages. Greater Jakarta contains approximately 8 per cent of the population of Indonesia and produces 11-14 per cent of the GDP.[6] Being the capital and a province some 100,000 people are employed by the central government and 30,000 work for the regional administration, hence the government sector plays a major role in the urban economy. By far the most important feature of the Jakarta economy is the dominance of the tertiary sector, particularly trade and services, which generate over half of the regional income. By virtue of the city's location, the volume of trade imported through Tanjung Priok, amounted to about 55 per cent of the total value of Indonesian imports in 1971.[7] While these factors may have lead to polarization and increases in the difference in regional income, a trickle-down machanism will eventually reassert itself when the spread of knowledge, transportation and standardization of all aspects of life permits further integration of the space-economy. Thus, according to the view of the modernizers, in the early stage of city primacy there will be increasing disparity between the developed and under-developed portions of urban space, but there is a tendency toward equalization as the economy of JABOTABEK reaches maturity.

This line of thought fails to impress the traditionalists, however. They argue that the inadequacies in society are a product of severe dis-economies of scale—the concentration of development and power in the Primate City. It is said that the primacy of the largest city frequently leads to the draining of vital resources which 'can only be reversed by a strategy of deliberate decentralization.[8] Radical changes in the nature of urban systems must be introduced to overcome the constraints of size, low growth rates, export orientation and colonial inheritance that confine modernizing influences to primate cities with little hope for downward filtering to take place. For instance, at present JABOTABEK is absorbing nearly 40 per cent of national private investments and over 15 per cent of the national development budget.[9] Yet, DKI Jakarta is short of development funds and the surrounding kabupatens of BOTABEK are beginning to realize their shortages of resources as well. A recent appraisal of the JABOTABEK economy indicates that the mean incomes are Rp. 76,000, Rp. 59,000 and Rp. 37,000 in Jakarta and the urban and rural areas of BOTABEK respectively.[10] For these and similar other reasons an increasing number of urban and regional planners begin to argue for programs of decentralization into new growth centers.

A decentralization pattern such as the one designed for JABOTABEK will probably cost less in infra-structure investment while avoiding many of the social perils of the supercity. Today there also appears to be a greater acceptance of the public sector and of international agencies in providing assistance to underdeveloped urban fringe areas in planning their economic growth. As to downward filtering, a new strategy may contribute less to the equities in society than to the advancement of further development using the growth center concept. But these may bring new and underutilized resources into the development process and also provide alternative magnets for migrants reducing the pressures on the Primate City. All of this should allow the natural filtering process to reassert itself as long as national growth and population control are successful. New regional structures can indeed overcome the constraints of the past and the limitations of the present. Both modernists and traditionalists appear to have been persuaded by many of these arguments. As a result, growth centers cum decentralization policies seem to have achieved greater status in the planning profession but evaluation must await the test of time and practice.[11]

Development strategies

The conflicting pressures in development strategies for JABOTABEK are formulated into two possible economic development scenarios : the "distributive" and "concentrated" strategies.[12] Within each of these approaches, key objectives (based on efficiency, equity and social criteria) for spatial and sectorial planning are identified.

Distributive Development : Under this strategy more investment would be channeled nationwide into agriculture and labor-intensive industries. Foreign investment should be reduced and concentrate only on technological and organizational uses. Public services such as education, health care, water supply and family planning would be expanded as rapidly as possible. Luxury housing and private passenger transportation would be discouraged. The "closed city policy" for Jakarta should be lifted and intensive efforts should get underway to create new employment opportunities. A distributive development approach should be coupled with increased contributions from the community and an improved tax collection procedure. It is anticipated that the rate of population growth in JABOTABEK will slow down and the overall income disparities could be much reduced.

Concentrated Development : At the national level major investments would go into large-scale manufacturing and other activities of the "modern sector"; as such JABOTABEK would continue to enjoy a high rate of growth. Agriculture, irrigation and small business services would receive a lower priority than in the distributive development technique. In addition, basic education, health and family planning targets would be moved back and generally less attention would be given to such issues as housing and transportation of the lower class. In contrast, middle income housing, road construction and private transportation are among high priority items in this development strategy. GDP growth in JABOTABEK would be considerably higher than under the distributive development method ; but considerable differences in income between Jakarta and the rest of JABOTABEK would remain.

Comparing the Two : The concentrated development strategy is more conservative in attempting to distribute income to the lower class. Also the GDP is relied upon to eliminate poverty, rather than governmental intervention. Clearly, this strategy would be easier to implement in administrative and political terms since it is more in accordance with the present distribution of income. On the other hand, it would not achieve the benefits of redistribution in income and human resources desirable for the future of JABOTABEK and the nation. In contrast, the distributive strategy aims at redistribution of income within the context of Pelita III and IV, and would directly therefore benefit the population of BOTABEK. This strategy will also more directly address the problems of BOTABEK, by emphasizing agricultural and service activities—rather than by activities controlled from Jakarta. In order for this strategy to become politically acceptable however, it would require strong leadership from the national government. Although it is believed that this approach could be implemented with present levels of resources, the benefits would be more widely distributed and thus be politically feasible.

Spatial alternatives

The JABOTABEK Development Plan tentatively identifies four spatial land use designs for tne JABOTABEK region which are based on the economic development scenarios discussed earlier.[13] In addition, these spatial plans must take into consideration changing land values, planning controls, land use integration and the like. For example, it should be noted that in the absence of effective development controls in Jakarta, investments in infrastructure will act as powerful forces in shaping urban design. Furthermore, land values have reached exceedingly high levels in the city in part because of land tenure conditions, the average lot size and a substantial excess of demand over supply of space. Another issue in determining spatial strategies is to recognize the causal relationships between economic development, employment population and land use. This relationship can be established by assigning a multiplier to population and land utilization based on the spatial allocation of types of employment. Finally, some emphasis should be placed on economic efficiency ; broadly defined as minimizing development costs and maximizing economic and social benefits. Some of this can be achieved by establishing incremental standards for development projects and by the willingness of users to pay for such standards (as in the construction industry). Clearly, these are some of the issues that diserve incorporation into the spatial planning options for they are particularly lacking in the BOTABEK area. Unless there is more effective land use control and guidance of major infrastructure to desired locations, planning strategies which depend on specific densities, size of holdings or building standards, are not likely to succeed.

There are clearly numerous options for the spatial allocation of the additional 13 million people projected to live in the JABOTABEK region by the year 2000. In the short-term, however, the options are limited to the continuation of existing development patterns. Beyond that, development costs and benefits are distinguished by whatever set of plans and policies are adopted ; hence, the need for long-range alternatives. The four alternative forms of urban growth outlines below have been proposed by various government agencies and are sufficiently different as to be amendable to financial and economic evaluation (Figure 13.2).

Alternative I : This approach envisions the continuation and intensification of current trends in urban development. DKI Jakarta will expand progressively as it does today, to the west, south and east in terms of employment opportunities and residential density. Assuming, of course, that such problems as the constraints on the aquifer to the south and the areas subject to flooding to the north-east can be overcome.

Alternative II : This option assumes that land use policies are set into motion to control the forms and functions of urban development. The plan is further characterized by a gradual improvement of existing facilities and the creation of several commercial sub-centers in order to improve services throughout the region.

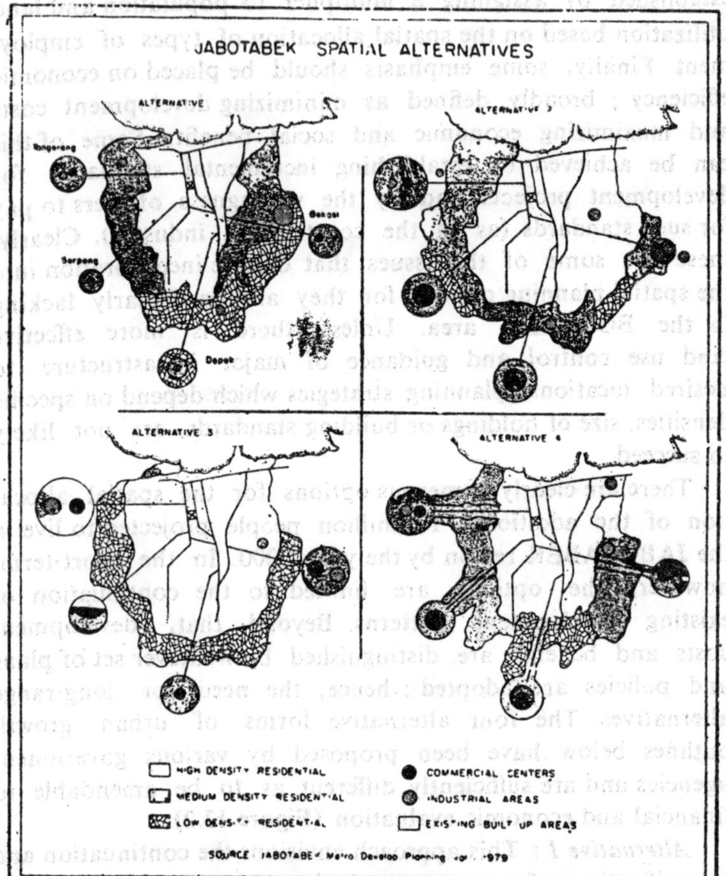

FIG. 13.2 : *JABOTABEK—Spatial Alternatives*

Alternative III : This plan resembles a modified growth center strategy. It assumes that strong controls are applied to encourage the development of relatively self-contained growth centers at Bekasi, Tangerang, Depok and Serpong. This is similar to the original JABOTABEK proposal which suggested that these centers should be expected to absorb larger populations and anticipate more growth in infrastructure.

Alternative IV : This proposal involves perhaps the strongest degree of input to control development. Its principal feature is the growth of corridors along major arteries between DKI Jakarta and the various regional centers. A similar proposal was made by Cipta Karya.

The JABOTABEK team is presently selecting the most appropriate scheme according to such criteria as maximum benefits, financial resources required, environmental impact, and institutional arrangements.

CONCLUSION

JABOTABEK is the first comprehensive government-sponsored approach to metropolitan planning in Indonesia. The principle objective is to identify target populations and development issues, assess their needs, while financial and administrative resources are determined. As the programs are carried out (during Pelita III and IV) the following improvements are expected to be realized : (*a*) increase agricultural and service employment for Botabek, (*b*) reduce current pressures on and overspill from DKI Jakarta, (*c*) balance the distribution of income and growth throughout the region, and (*d*) establish a guided land development scheme for the next 20 years as to control land prices, orderly growth and supply more land for low income settlements.

Financed jointly by the Government of Indonesia and the World Bank, the project is action-oriented. It is intended that the end product will better reflect the concerns of government and have a greater probability of being implemented than most consultant's recommendation which may or may not be carried out. Yet, the present JABOTABEK concept should be considered an ongoing planning process at least until the turn of the century. It is indeed only a framework for more detailed planning by various jurisdictions and agencies responsible for providing infrastructure and services.

FOOTNOTES

1. King, D.Y. and Weldon, P.D., "Income Distribution and Levels of Living in Java 1963-1970, *Ekonomi dan Keuangan Indonesia*, Vol. 23, No. 4, 1975, p. 375.

2. *Jabotabek Metropolitan Development Planning*, Vol. II, 1979, p. 187 (Ministry of Public Works ; Directorate General of Housing, Building, Planning and Urban Development ; Directorate of Urban and Regional Planning).
3. *Jabotabek Metropolitan Development Planning—Proposed Guided Land Development Program for Greater Jakarta* (Summary), June 1980, p. 2.
4. *Jabotabek Metropolitan Development Planning*, Vol. I, 1979, p. 28.
5. *Jabotabek Metropolitan Development Planning*, Vol. I, 1979, pp. 4-5.
6. *Jabotabek Metropolitan Development Planning*. Vol. I, 1979. p. 1.
7. Sethuraman, S.V., *Jakarta—Urban Development and Employment*, International Labor Office, Geneva, 1976, p. 2.
8. Berry, B.J.L. "City Size and Economic Development : Conceptual Synthesis and Policy Problems, with Special Reference to South and Southeast Asia," in : *Urbanization and National Development*, editor, Jakobson, L. and Prakash V. Sage Publication, Beverly Hills (Vol. I), 1971, p. 112.
9. *Jabotabek Metropolitan Development Planning*, Vol. II, 1979, p. 157.
10. *Jabotabek Metropolitan Development Planning*, Vol. I, 1979, p. 21.
11. Berry, B.J.L., 1971, p. 143.
12. *Jabotabek Metropolitan Development Planning*, Vol. I, 1979, pp. 34-38.
13. *Jabotabek Metropolitan Development Planning*, Vol. III, Chapter 8-9.

FOURTEEN

PRITHVISH NAG

LUSAKA
Problems in Developing an African Capital City

LUSAKA, a capital city of south-central Africa, seems to be most complicated urban centre of Zambia. The various approaches in planning, complicated land holdings, unplanned settlements, statutory and unstatutory development, gaps between the built-up areas, and social and economic segregation of different quarters have made urban development process for the city an intricate exercise. Geology, especially dolomite, quarzite, schists rocks and high water table in rainy season along with the karstic terrain have further complicated its planning.[1] The impulses of growth and decay of Livingstone town has also been experienced here, but, not at the same wavelength.[2] Lusaka's planned and unplanned development have shaped the present morphology of this city (Fig. 14.1). Keeping in view of these problems it has been chosen so as to study the most matured problems confronting the planning of the urban settlements of Zambia.

Lusaka developed from a small African village at the railway halt (1904 A.D.) to the largest urban centre of Zambia. It became a multi-functional settlement by adding agricultural, transportation, administrative, industrial and commercial functions.[3] With the multiplicity of these functions, planning became necessary. In 1905, a grid-iron plan was prepared for this settlement near the present railway station. This area now forms the City Centre of Lusaka. In 1928 the planned area was still undeveloped and there were four sub-divided farms known as *Emmasdale*, *Villa Elisabetta*, *Fairview* and a part of Farm 90a.[4]

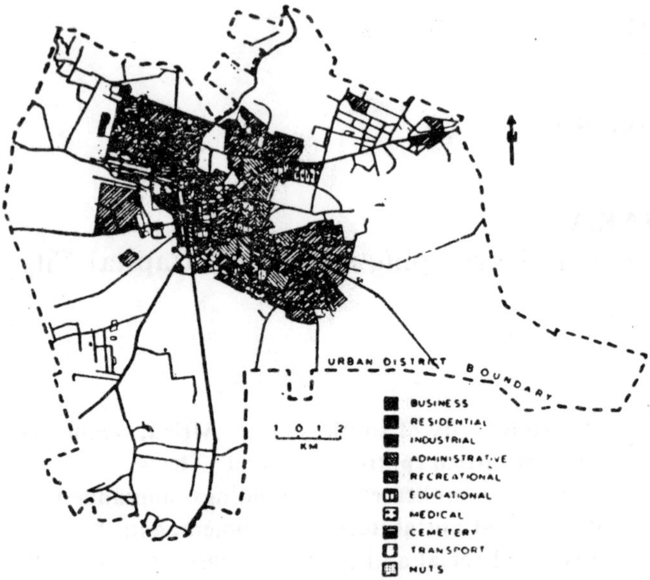

FIG. 14.1 : *Lusaka—Landuse*

During this period the territorial capital was at Livingstone. The colonial government engaged S.D. Adshead and later P.J. Bowling to select a new site and to plan this capital city. The settlements developing slowly on the grid-iron pattern was finally selected for a 'generous-gracious-garden city' making the full use of the ridge (now known as *Ridgeway*) lying east of the line-of-rail. Collins refers to the non-statutory development plan which started in 1938.[5] This was replaced by the first statutory development in 1952, which was the year when the settlement was raised to a Municipal Council having proper planning authority.

The implementation of the Bowling plan brought in two other nuclei *viz. Ridgeway* and *Government House* (now *State House*), lying east of the old town. Plot size and land values were planned keeping inview of the residential requirements. This ultimately developed into a linearly arranged multi-nuclei settlement. The space in between and around these nuclei are being occupied by the built-up area. Figure 14.1 shows the rapid change in the built-up areas after 1962.[7] The new areas, such as the university complex, are being added to the city.

At the time of independence this settlement added many legacies and problems to town planning. The myth of a garden city was in opposition to the growing needs for housing. On the other hand, *Olampia Park* (now *Chikonkoto* within, and *Roma* and *Kabulonga* outside the council boundary were private European housing areas. *Handsworth Park* (now *Shikoswe*) and the *Chama Valley* small holdings remained as European housing complex in the north eastern side of the town.

Indians were allowed to reside behind their shops and later in *Luburma* area. Indians have built a *Hindu Hall*, *Mosque* and *Gurudwara* (*Sikh temple*) in Lusaka. Indian houses are also scattered throughout the city and even in the suburban small-holdings such as *Makeni*. In addition, there is a sizeable Indian expatriate population in Lusaka.[8]

The industrial areas have developed according to the plan. There are two industrial areas : (a) east of the line-of-rail, and (b) south of the *Matero* housing area. Both these sites are linked with the railway line.

The problem of low cost housing has always complicated the planning of Lusaka City. *Kamwala* and *Chibolya* which were included in the Bowling Plan (1933) were not sufficient to accommodate Lusaka's African population, especially after 1948 urban Housing Ordinance when for the first time a provision for housing for every employed African within the council boundary was made. This ordinance helped in developing African quarters such as *Kabwata* (before 1945), *Chinika* (1947), *New Chilenji* (1950), *Chiebolya* (1951) and *Matero* (1952).[9] The other African housing areas also developed along the sewage line even after the enforcement of statutory plans. These high density areas are *New Kamwala* (1962), *Libaba I and II* (before 1962) and *Libaba III and IV* and *Chilenje South State I and II* (1965).

It was soon felt that high density areas for Africans were not sufficient for the growing population. Unplanned housing areas or squatter settlements developed surrounding the town. *Kalingalinga*, *Quarries*, *Marrapodi*, Kanyama and George are the colonial unplanned housing areas. They accounted for about 10 per cent of 1963 population.[10] However, there are a large number of villages around Lusaka having similar characteristics to that of squatter settlements. Around 1966 the existing squatter settlements expanded in all directions, while the new settlements sprang up in the intervening vacant spaces (Fig. 14.2).

According to the Census of 1969, Lusaka city had 35,659 dwelling units out of which 68 per cent were of *pole and dagga* walls, 42 per cent of asbestos roofs, and 38 per cent of corrugated iron roofs. This clearly indicates the temporary nature of residential quarters in Lusaka. About 1973-74, the unplanned housing areas further expanded and smaller ones were included to form a big unplanned settlement area. Two most significant development may be mentioned here. First example is regarding the conversion of *Mandevu* and *Marnapodi* (*formerly Hillside*) unplanned areas into a planned residential area, known as *Mutame*. The other example is of *Kaunda Square* in the north of the *Great East Road* on self-help basis. Such development was possible with the help of the Government and the Lusaka City Council.

The problems of Lusaka is not only the co-existence of planned and unplanned settlement areas, but there are several other problems which come in the way of development. These are : (a) traffic congestion at certain points, (b) odd location of the city centre, (c) poor links between the housing areas, (d) complicated land rights, and (e) existence of farmsteads and small-holdings around the city.

Lusaka City Council has engaged Doxiadis Associates for the preparation of a development plan.[11] This capital city has a bright future in south-central Africa and will continue to attract population and development. It has proved to be a growth pole for Zambia. Proper planning is necessary keeping in view of its national importance as a highest order service centre.

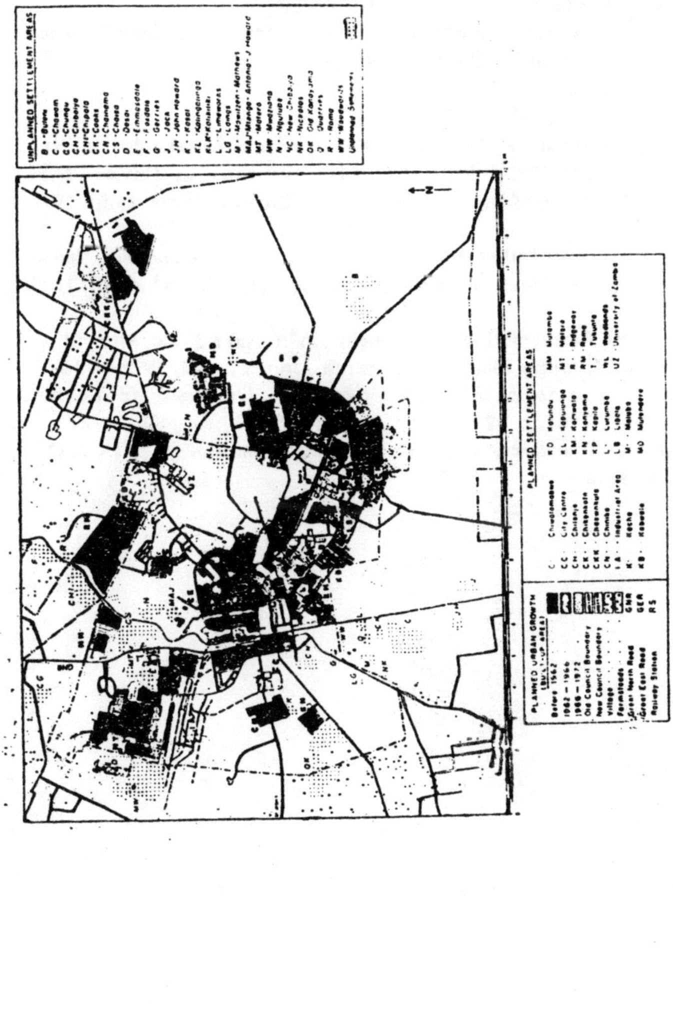

FIG. 14.2 : *Lusaka : Capital City*

FOOTNOTES

1. Davies, D.H., 'Lusaka, Zambia : Some Town Planning Problems in an African Capital City at Independence', *Zambia Urban Studies*, University of Zambia, 1973, pp. 4-5.
2. Nag, P., "Basis of Zambian Settlements", *Africa Quarterly, Indian Council for Cultural Relations*, Vol. XVII, July 1977, pp. 59-76.
3. Lusaka gained the status of Village Management Board in 1913, Municipal Board in 1953, and City in 1960.
4. Collins, J., 'Lusaka : The Myth of the Garden City', *Zambia Urban Studies*, University of Zambia, 1969, p. 4.
5. Collins, *idem.*
6. Davies, *op. cit.*, 1973, p. 7.
7. For the growth of planned built-up area of Lusaka during 1913, 1929, 1933 and 1952 vide Davies, D.H., "Lusaka" in Davies, D.H. (Ed.), '*Zambia in Maps*', University of London Press Ltd., 1971, p. 87.
8. Nag, P., "Asian Settlement in Zambia", *Africa Quarterly, Indian Council for Cultural Relations*, Vol. XVIII, No. 4, April 1979, pp. 59-77.
9. 'Low Cost Residential Development in Lusaka', Department of Town and Country Planning, Research Unit, Lusaka, 1972, p. 10.
10. Kay, G., '*A Social Geography of Zambia*', University of London Press Ltd., 1967, p. 127.
11. Doxiadis Associates, *Lusaka : Existing Conditions*', DOX-ZAM-A38, Lusaka.

FIFTEEN

HARENDRA PAL SINGH

METROPOLITAN PLANNING WITHIN THE FRAMEWORK OF REGIONAL PLANNING IN DEVELOPING WORLD

THE first question in metropolitan planning is "What can be done to master urban growth ?" The only answer of this question is in one word—Planning. It is the beginning and end of the *communis opinio doctorum.*[1] The specific questions related with time horizon, spatial perspective, institutional barriers, participation, information systems, research priorities etc. are before planners first to enter in metropolitan planning.

Time horizon is not fixed for metropolitan planning. It varies in different conditions and situations to solve the leading metropolitan problems. Nevertheless we should try our best to find some common denominators for clear definition of the range of choices.

It can be divided into three motions : (*i*) short-term intervention ; (*ii*) medium-term planning ; and (*iii*) long-term prognosis. The key issues of metropolitan planning should be solved within the framework of medium-term planning which ranges between 5-15 years.

This time horizon is long enough for structural changes to create favourable conditions in a metropolitan region but it is short enough to anticipate about population, employment, investment, consumption, economic social, technological and institutional innovations, landuse patterns etc. The short-term intervention may be applied if the choices in planning are limited.

In metropolitan growth, planning means less intervention. Medium-term planning may be applied in the areas where the strategic choices are relevant. Short-term planning is supplementary to medium term planning. It adjusts in the current situations and coordinates all actions in the implementation stage.

Medium term planning needs long-term counterparts also. The development of futurology is disclosing new possibilities for the improvement of medium term planning. Futurology may be defined as an *avant garde* of planning. We should not pretend to know the future as it is beyond our intellectual capacities.

The time horizon should not be extended beyond the year 2000 as the time of 20 years is long enough to test our different hypothesis on the future of the society. Of course, in the domain of imagination and fantasy we can discuss the problems of the twenty first century. There is a difference between these two types of futurology : (*i*) *as an avant garde* of planning—a field of activities to improve planning and decision making, (*ii*) a futurology based on the domain of imagination and fantasy with no relations of the world and no limits of time and space.

We can explain it with the help of colloquium of J.G. Papaioannou :[3]

"If one takes the configuration of Ecumenopolis which represents one of its more probable models, these primary centres are seven in number, according to some simple geometrical considerations concerning regular poly-hedrons inscribed in or circumscribed around the earth, as well as other considerations on the economy in the hierarchical layout of settlements (but entirely unrelated to present political conditions). In the northern hemisphere they form a sort of belt in middle latitudes with the main centre among the four larger ones being the one of Western Europe (London, Paris, Randstadt). Next in importance come the two centres on the eastern edges of the two main continents, *i e.*, a centre close to the East Coast of China around Shanghai, with Tokyo as a twine one. A fourth in the northern hemisphere is in California. In the southern hemisphere, again in middle latitudes, we find three main centres : One in South America, around Buenos Aires, one in South Africa, and one in South-East Australia, in the Melbourne-Sydney area. These seven centres, with a probable eighth in the Antarctica as a still more remote future time (when Antarctica will have been more fully conquered in its habitability) constitute the main focal points on which urban development around the earth will depend. A system of secondary and then tertiary, quaternary etc. centres can be developed around these primary ones, all of them inter-connected by urbanization axes of a corresponding hierarchical level, the highest order axes being those connected with the highest order centres and so on."

The micro, the macro and the mega approaches should be distinguished in the *spatial perspective* of metropolitan planning. Dealing metropolitan region as an autonomous phenomenon, the *micro approach* is similar to urban and physical planning. There are three main weaknesses of this approach :

(*i*) Though social and economic goals are significant, the approach is very much concerned with physical objects to develop transportation system.

(*ii*) The metropolitan planning offices are not the part of metropolitan government. They are treated as the study group, the proposals of which will or will not be implemented by the local government.

(*iii*) This type of planning is very loosely related with national policies and plans. The main intellectual and managerial efforts are being made to integrate the local planning units in the metropolitan region.

The macro approach is of the view-point that the planning of metropolitan region may be more successful in a nationally integrated system of planning and decision-making. This approach applied theories and models developed in social and economic planning at the national level in metropolitan planning. Hence, the change from micro to macro is a substantial change in metropolitan planning. The classical role of landuse planning is also considered in metropolitan planning but it is not involved in leading social and economic problems of the metropolitan region. The metropolitan planning is an integral part of national planning. The system of national planning includes overall planning ; sectoral planning ; and regional planning.[4] There are two main functions of this system in metropolitan planning. It supplies information of the development of the national economy and society, sectoral development, and the development of other metropolitan and non-metropolitan regions of the country. It also creates external consistency check of the plan for the metropolitan region. The feedback relationship exists between the metropolitan planning and the national planning. It not only uses the informations of other plans but also provides new information to overall sectoral and regional plans.

In fact, the macro approach supplies the most effective perspective of metropolitan planning but it dies not mean that it should not be supplemented by micro and mega approaches. The spatial perspective of micro approach is shifting from the scale of region to the scale of the local community and of *mega approach* from the nation to the continent. The mega perspective is an *avant garde* of macro planning or a domain of fantasy. It indicates the limitations in metropolitan planning in national boundaries and provides positive suggestions to diminish the negative consequences of the phenomena. It assumes non-existence of political boundaries and the national states.

Metropolitan planning is a comprehensive phenomenon where interaction and readjustment can be seen in goal setting, plan design, plan implementation and plan evaluation.[5] Hence mere plan for a metropolitan region with splendid cartographic annexes is not enough for a study group.

Implementation is a real test of the performance of planning. However, most of the planning bodies in several countries exert little influence on the implementation stage. The sectoral and local institutions dominate over this stage. Hence the changes at institutional level are important for efficient metropolitan planning.[6] New administrative divisions should be introduced and historical or medieval boundaries should be eliminated. They provide efficient barrier for modern policy and planning. The metropolitan region should form the core element of any new administrative division. The new administrative units which shall be equipped with administrative strong government is the background for a metropolitan planning.

Planning office often appears to be professional technocratic ivory tower, working in a splendid isolation related to the given society and community. The metropolitan planning should be based on understanding of human beings, on promoting a dialogue with the citizens and on participation in the planning process. The creation of efficient participation is a challenging task of social sciences.

The quality of metropolitan planning depends upon the quality of information which the planning process is able to use, process and diffuse. The metropolitan planning offices are processing scattered and disintegrated informations. Recently efforts have been made on international level to build and develop information systems for regional planning.[7] The conclusion came as the metropolitan region did not need only planning office but a data bank also for necessary information regarding efficient planning and administration.

Metropolitan planning : case studies

Delhi : The Delhi metropolitan area of about 1280 sq. km. comprises the area of Union Territory of Delhi and the ring towns of Loni, Ghaziabad, Faridabad, Ballabgarh, Bahadurgarh, Gurgaon and Narela. To achieve a rational growth of Delhi, it is essential to plan this whole area as a composite unit for integrated and balanced development. The area is homogeneous with physical, social, economic and cultural unity. The right towns should be developed in a planned way to check the migration from this area to Delhi. It is the only way to prevent urban sprawl which is threatening to grow into one giant urban mass.

The development of the right towns can be ensured by strengthening industrial structure and locating government offices in these towns. In 1951 the total populatiom of these six ring towns was 1,25,000 and the projected population for 1981 is 8,42,000 (Table 15.1). The total requirement of land is 21,300 acres. The projected industrial employment for 1981 is 1,14,000 and the land requirement for industries is estimated at 4,430 acres. Similarly, the employment in government offices for 1981 is 50,000 for which 230 acres are required[8] (Fig. 15.1).

Ghaziabad : The U.P. Govt. has prepared a separate plan for Ghaziabad. It is envisaged as an industrial town with 50,000 workers engaged in manufacturing out of a total population of

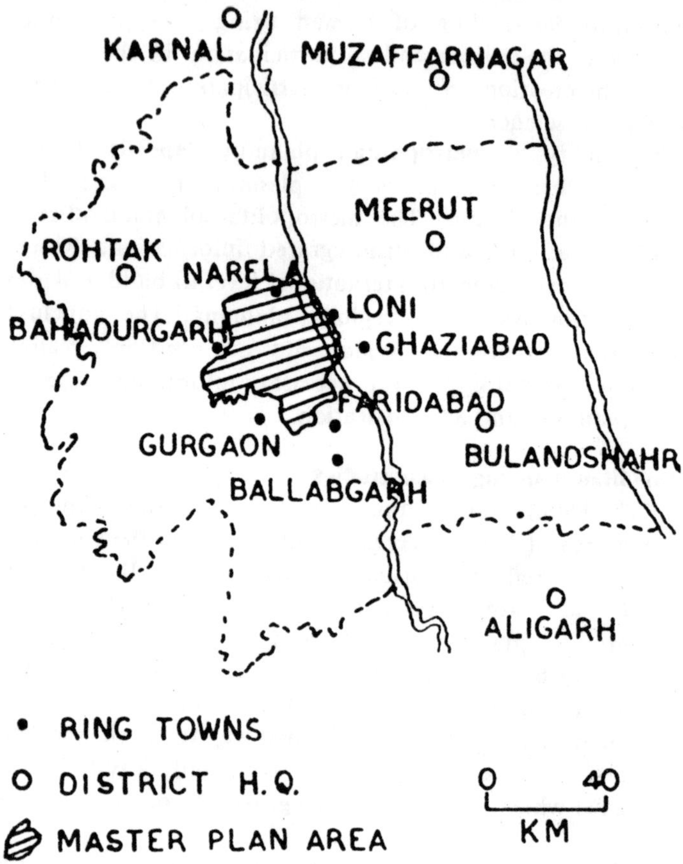

FIG. 15.1 : *Delhi Metropolitan Region*

about 3,57,000 in 1981. 70,0:0 people will get employment in government offices. The land under industries and government offices shall be 2,000 acres and 200 acres. The residential, commercial, parks and other public and semi-public we will cover about 8,000 acres.

Faridabad-Ballabgarh : Ballabgarh, and the new and old town of Faridabad are jointly situated. The Mathura road runs through the area along which industries are springing up like mushrooms. A master plan is also functioning in the area by the Haryana government. Like Ghaziabad, Faridabad also

TABLE 15.1 : **Land requirements in ring towns·1981**

Ring Towns	Population in 1000's		Manufacturing Employment and Land 1981			Government Employment and Land 1981		
	1951	1981	Total workers	Employment density per acre	Land acres	Total workers	Employment density per acre	Land acres
1. Ghaziabad	44	357	50,000	25	2,000	20,000	200	100
2. Faridabad	31	285	40,000	30	1,330	15,000	250	60
Ballabhgarh	6	29	3,000	20	150	—	—	—
3. Gurgaon	19	48	5,000	20	250	5,000	250	20
4. Bahadurgah	11	29	4,000	20	200	—	—	—
5. Loni	4	23	2,000	20	100	—	—	—
6. Narela	10	71	10,000	25	400	10,000	200	50
Total	125	842	114,000	—	4,430	50,000	—	230

Source : Delhi Master Plan 1962.

record industrial base with about 43,000 workers engaged in manufacturing out of 3,14,000 population in 1981. The government offices will accommodate 15,000 employees. The land requirement will be 1,500 acres in manufacturing and 60 acres for government offices.

Gurgaon : This town is handicapped for want of good water resources. 5,000 industrial workers have been proposed out of 48,000 population in 1981. The industrial estate will cover 250 acres and another 20 acres to locate government offices with 5,000 workers.

Bahadurgarh : This town also does not record good sources of water supply. It will record 29,000 population in 1981. About 200 acres would be required for industrial employment for about 4,000 workers.

Loni : This small village town of 4,000 population is to be developed for 23,000 population in 1981. An industrial base of 100 acres with 2,000 workers will provide the economic base.

Narela : This township is proposed to be developed to absorb 60,000 people by 1981 in addition to its existing 11,000. It is situated under Delhi Master Plan.

Landuse plan for urban Delhi

The landuse plan for urban Delhi is conceived to find solutions for the entire complex of inter-related urban problems, *viz.*, social, economic and governmental. It may be called a comprehensive plan as it aims at balanced and integrated development to take care of future growth up to 1981. One of the major policy decisions, *viz.*, of considering urban Delhi as the core of the larger Metropolitan complex and the necessity of developing the Ring towns with a sound industrial base and employment in government offices.

The following table gives the total account of area in different landuse categories (Table 15.2).

Residential areas

The main objective of the plan is to relate residential areas with government offices, industrial and commercial areas to minimize the journey to work and back. With this viewpoint, a rational pattern of residential densities is proposed for correcting the present disparities and imbalances. Higher densities are

TABLE 15.2 : **Proposed landuses in Delhi metropolitan area**

Landuse	Area in acres	Per cent
1. Residential	48,638	40.2
2. Commercial	1,615	1.3
3. Industrial	8,109	6.7
4. Governmental	1,149	0.9
5. Defence	8,638	7.1
6. Recreational	27,342	22.6
7. Public and Semi Public Use	9,282	7.7
8. Railways	2,251	1.9
9. Airport and Transport Facilities	5,175	4.3
10. Agricultural	8,805	7.3
Total	1,21,004	100.0

planned nearer to the work areas. Lower densities are proposed towards the outer limits of the city.

The old city of Shahjahanabad records a gross density of about 350 persons per acre leading to unhealthy living conditions and functional obsolescence. Vacant plots and dilapidated structures will be acquired for community facilities like schools, parks, health centres etc., for a gross residential density of 250 persons per acre. New Darya Ganj which has been developed at a density of 200 persons per acre needs conservation measure. They are also required in Malkaganj, Aryapura, Sadar Bazar, Nanak pura, Motia Khan, Qadam Sharif and Pahar Ganj. Mata Sundari, Sarai Rohilla, and Shadipur-Khampur are sub-standard areas. They should be developed at 200 persons per acre.

Almost all New Delhi residential areas and others are low residential varying from 50 to 100 persons per acre. They too need lccation of community facilities. The whole area for population and residence has been divided into eight divisions :

Commercial areas

Chandni Chowk, Khari Baoli and Sadar Bazar form the Central Business District (CBD) of the metropolis. General business and commercial activities are seen in Nai Sarak, Chawri Bazar, Kashmere Gate, Faiz Bazar and Asaf Ali Road. The redevelop-

TABLE 15.3

Planning divisions	Proposed 1981 population	Residential area (acres)
1. Old City	3,22,600	1,370
2. Civil Extension	3,98,200	2,590
3. Civil Lines	3,55,200	3,480
4. New Delhi	6,34,100	6,930
5. Shahdara	7,44,100	7,890
6. South Delhi	7,81,100	9,400
7. West Delhi	7,44,300	8,240
8. North West or West Yamuna Canal	6,06,200	7,460
Total	45,85,800	47,360

ment of these areas will obtain the supreme position in the commercial activity of the Delhi Metropolitan area.

Connaught Place is another CBD of the capital. Commercial development is also observed along Parliament Street. Further commercial activities are planned in the Keeling Road area. Minto Road, Ranjit Singh Road, Kamla Market, Ram Lila Grounds, Delhi Gate and Ajmere Gate are planned for intensive commercial activities to function as the lung for the Old City. Other sub CBD are seen along Ajmal Khan Road, Ghaffar Market in Karol Bagh and Shahdara.

In order to decentralize commercial activity, the metropolis has been divided into 15 District centres.

Industrial areas

The working force in urban Delhi in 1981 is estimated at 4,40,300 and 1,14,700 in the ring towns of the Delhi metropolitan area. There are two planned industries districts in Delhi. One is along the Najafgarh road which records both extensive type of manufacturing units like D.C.M. chemicals and intensive industries like Hindustan Breakfast Food Factory. The other is a small scale industrial estate near Okhla. Rest of the industries are scattered all over the city.

Flatted factories are proposed in Ajmere Gate, Thompson Road, Motia Khan, G.B. Road and other neighbouring areas in order to provide work centres within walking distance of the existing population. Extensive industries are proposed in South of Okhla, Najafgarh road and Shahdara.

Governmental areas

About 2,62,500 persons shall be employed in government offices by 1981 in urban Delhi and about 50,000 in the ring towns of the Delhi metropolitan area. At present the Central Secretariat Complex and Parliament Street are the centre of Government offices. Indraprastha Estate on Mathura Road and Old Secretariat in Civil Lines are subsidiary centres. Other government offices are located in Shahjahan Road and scattered areas.

Loni Road, West of Safdarjang Hospital, Ring Road District Centre, Kalkajee District Centre, North and South Shahdara, West Delhi District, New Civil Centre on Circular Road, Chanakyapuri and Shalimar Gardens are the planned sites for further extension of governmental areas.

Recreational areas

A system of parks has been worked out for the entire urban area of Delhi related to residential densities. Local parks and playgrounds range from 2.00 to 0.81 acres per thousand population for residential densities varying from 25 to 200 persons per acre. The district parks are proposed on the basis of 3 acres per thousand population and each district centre has a district park nearly. The total area earmarked for district parks is about 17,940 acres including semi-public recreational areas.

The existing parks are the Parade Grounds, Rajghat, Queen's Gardens, Ajmal Khan Park, Anand Parbat, Roshanara Gardens, Central Vista, The Hexagon, Lodi Gardens, Ramlila Grounds, Talkatora Gardens, Vijay Chowk, Zoological Gardens, Buddha Gardens and Lady Hardinge Ground. Several other parks are proposed in the master plan of Delhi.

Public and semi public uses

There were 26 colleges affiliated to Delhi University in 1960. 22 Colleges are proposed in the same University. In addition to Delhi University, J.N. University, Jamia Milia Islamia, I.I.T., NCERT and several other educational institutions are functioning within the capital. There are at present 7 general hospitals and a number of private hospitals and maternity hospitals. The Government has proposed 20 more hospitals in the city.

Several other departments related to police, water, power, sewerage, wireless and radio, cremation, dumping grounds and sanitary fills and social-cultural institutes are functioning and are proposed in the plan.

Traffic and transportation

In order to bypass Delhi Main Railway Station, the Railways are planning a goods avoiding line as a ring railway for local suburban traffic. Okhla, Nizamuddin, Subjimandi, Azadpur, Baoli, Shahdara, Shakurbasti and Nagloi Railway stations are under improvement and extension. The Marshalling yard at the New Delhi Railway Station will be shifted to Tughlakabad on the Agra line.

In order to avoid congestion in the urban core, several ring roads have been constructed and planned linking them with the inner ring roads.

Karachi[9]

Karachi is the largest Metropolitan area of Pakistan comprising significant and vital industrial and commercial characteristics. In 1967, it recorded 40 per cent of industrial production of West Pakistan. During this period the metropolis consumed 50 per cent of the total raw material and spare parts. Out of 35 banks in the country, 26 are located in Karachi. It is estimated that about 53,000 commercial and service establishments are located here. It has a total population today of more than 3 million. It is the only port serving Pakistan and landlocked Afghanistan.

Pakistan had a total population of 17 million in 1970. Out of this Karachi records 38,83,677 persons. Based on 1961 census 2.1 per cent of the total population were industrial workers, 656 per cent were total civilian workers and 16.7 per cent were non-agricultural workers in Pakistan. 13 per cent of these industrial workers were in Karachi metropolitan area (Fig. 15.2).

Historical development of Karachi

The period from 1947 onwards, the city faced several unexpected demands and developments. Karachi was made the interim

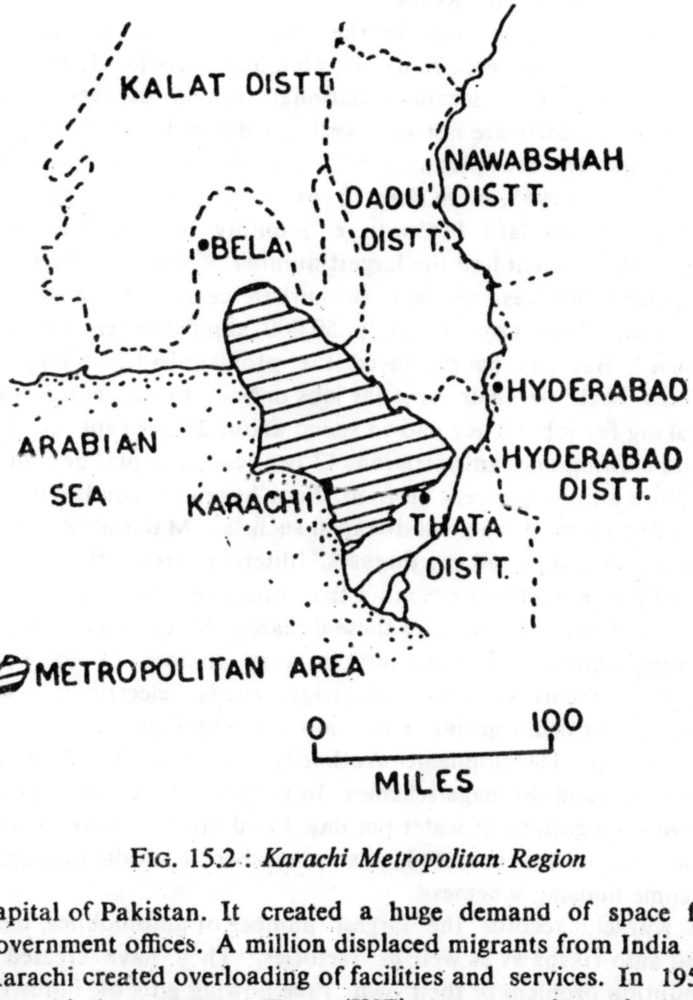

FIG. 15.2 : *Karachi Metropolitan Region*

capital of Pakistan. It created a huge demand of space for government offices. A million displaced migrants from India to Karachi created overloading of facilities and services. In 1952, Karachi Improvement Trust (KIT) was created to solve these problems. The KIT prepared the master plan for Karachi. The major attention in the plan was given for slum clearance and low income housing, provision of amenities and services for lower income groups, and general planning and development of the metropolis. Rapid growth of industries and commerce created shocking slums, serious transport and housing problems, mounting shortages of water, and an overall chaotic development pattern of the whole metropolis.

Current problems and issues

The basic problem lies in the lesser integration of social and physical planning with economic planning at national, regional, and local levels. Secondly national, provincial, regional and local level efforts are not very well coordinated and integrated. Consequently the planning efforts could not solve the slum problem. In 1965, more than 72 per cent urban people were living sub-standard life with a monthly income of less than Rs. 200. Karachi had the largest number of slum dwellers and displaced families. A new sub urban centre (Korangi) was planned where approximately 50,000 slum dwellers were re-housed. But the people faced the problem to travel long distances back to their original jobs or back to the central areas looking for jobs. They had to spend about 20 per cent of their total earning on transportation, 13 per cent on rental and other utilities and 50 per cent on food. Poor hygeinic conditions and bad diet resulted seasonal diseases, such as Malaria, dysentry, typhus, and skin and eye diseases. Illiteracy rates vary from 40 to 60 per cent. There are more than thirty resident "pirs" and several "hakims" to cure these diseases. Modern medical care is very limited. The community is very anxious to solve the basic problems of water, drainage, roads, electricity, waste disposal, health, housing, education and employment.

Karachi Development Authority contributed the water, sewerage, and drainage schemes. In the first phase, it supplied 70 million gallons of water per day. In addition to lower income schemes, it has executed a large number of middle and upper income housing schemes.

Karachi records the largest number of automobiles, buses and auto-rikshaws as well as factories. They have created a pollution problem of their own. Free flowing gifts of industrialization and development naturally affect the health of all the residents. Most of the planners and researchers are ignorant about these problems.

The United Nations and the Dutch projects

The major planning efforts in Karachi initiated in 1970 under United Nations auspices. In the same year Netherlands Government also took initiative on slum improvement and urban development problems in Karachi. The projects were aimed to provide the answer to the questions, what, why, where, and when to invest for harmonious socio-economic, functional, and spatial development of the region.

The Master Plan Development (MPD) in Karachi has established a number of working units such as a unit for regional planning and development, an urban design unit, an economic and legal unit, a social analysis and social planning unit, a housing and utilities unit, a transportation unit etc. These units are taking preliminary research work.

The Dutch Project concluded that major problems of this slum settlement were : (1) employment and related economic issues ; (2) basic utilities and services (water, roads, drainage, electricity etc.) ; medical care and health education ; (4) education ; (5) social planning and development ; (6) community leadership ; and (7) development of an integrated community.

Future directions

Jagdish Bhagwati[10] exposed the shortcoming of the national plans and finds that neglect of spatial dimension are responsible for the failure. There is a scramble by regions, metropolitan areas, and their political leaders try to grab industrial projects generating all kinds of inefficiencies, not only in terms of economic planning but also in terms of spatial planning and organization because the areas which get these projects are not ready for such growth, rational or irrational.

In addition to spatial component of the national planning efforts, communications network between national, regional and local level of planning is essential. There is a general consensus that the traditional growth models of economic development should be replaced by a more integrated approach for economic, social and physical development.

Clear urban land policies as well as urban land reforms are prerequisite of metropolitan plans. New institutional, financial, and legal arrangements are essential before implementation. Planners cannot arrange these prerequisites hence national plan needs to be converted into political action for basic institutional reforms and changes.

Calcutta

Calcutta metropolis is significant as the home of about nine million people ; it is significant as state capital and only important city of West Bengal with its 45 million population, it is important as the primate city of Eastern India of 150 million people, and it is significant as one of the three national cities of the world's largest democracy. It is larger than 85 independent nations and its hinterland would rank fifth in the world in population.[11] In terms of its complex mixture of the primitive and sophisticated, in terms of its population pressure and continuing deterioration of urban life Calcutta represents the leading edge in the developing world. It is in difficult straits such as deficiencies in housing and public services ; a deteriorating quality of urban life ; precarious to disastrous states of local fisc ; increasing unemployment with an increasingly restless and vocal educated segment.

The context

There has now been an extensive and heavily manned effort to devise plans and programmes to alleviate current problems and to turn Calcutta on gradual improvement. The aspects peculiar to Calcutta metropolitan planning are the character of population pressure ; the character of the metropolitan economy ; the relationship between Calcutta and its hinterland ; and the Bengal political situation.

The chief characteristics of the efforts

The planning efforts in Calcutta may be summarised as :

(1) The efforts are very large in terms of finance and manpower ;

(2) The efforts are made in the larger regional context of Eastern India ;

(3) The planning efforts are in remarkably broad and inclusive terms ;

(4) The Development Plan of Calcutta metropolis was policy-oriented, dynamic, options-open plan, different from conventional metropolitan plans ;

(5) The plan has set standards and devised programmes according to the resources of the metropolitan district ;

(6) The plan has priority for those areas which can develop the area toward its chosen goals and objectives and rejects those areas which may inhibit economic development ;

(7) The plan ends with an immediate action Development programme to make the city livable.

Size and continuity of efforts

The Calcutta Metropolitan Planning Organization (CMPO), the biggest in India, had 600 employees. The annual budget of CMPO is Rs. 3,500,000 in which ford foundation has spent $5 million in various instalments. Later on to coordinate the CMPO, Calcutta Metropolitan Development Authority (CMDA) was established with 800 employees. The project designs and programming fall within the purview of the CMDA, whereas systems planning, long range planning and state planning function with the CMPO.

The regional character of planning

The population of Calcutta metropolitan district in 1971 was 4.3 million. The projected population in 1986 will be 12 and 13 million if there is no change in the present position of Calcutta in relation to other urban centres. The population of Eastern India in 1961 was 111 million which is expected as 191 million in 1986, a staggering increase of 80 million. Eastern India is least urbanized as it records only 13.1 per cent urban population. Like other urban plans, CMPO devote space to consideration of the city's hinterland.

The breadth of the planning efforts

The planning efforts included, projected system plans up to the year 2001 for water, sanitation, and drainage ; for transportation plan up to 1986, for the improvement of 3,000 bustees ; for economic development, for governmental reorganisation in the metropolitan district ; for housing, education, and health, for immediate action programmes, and for future form of generalized land use plan.

There are three areas in which work has been less successful. They include economic development planning, planning for education and health, and housing. There are a number of reasons for this. First, economic recession of 1967 in India, hit Calcutta hardest. And, though India recovered it but recovery in West Bengal never gained momentum. The period of 1966-70 showed a decline in real incomes and an absolute decline in

TABLE 15.4 : **Registered factory employment in different industry, West Bengal**

(hundreds of thousands)

Industry Group	1951	1961	1965	1969	1970
Engineering	1.58	2.50	3.24	2.95	2.94
Jute	2.78	2.02	2.59	2.04	2.25
Others	2.16	2.66	2.97	2.92	2.90
Total	6.52	7.18	8.80	7.91	8.09

employment. This picture is more clear from the following Table.

The plan had the belated beginning on a long-range spatial plan for education facilities for the metropolitan district. The CMPO is undertaking a long range plan for health facilities and health services. This plan will include curative medicines, preventive medicine, nutrition, maternal and child health care, family planning, and health education as an integrated package.

Bustee Improvement Programme is one of the successful area of planning and programming. A bustee is the hut slum of Calcutta similar to the favela of Rio de Janeiro. In the Calcutta metropolitan region at least two million people live in such settlements. Their size varies from a few thousand people to 50,000 people. The programmes of introducing water lines and water taps, sanitary latrines and sewers, paving of lanes, street lighting, improvement of small open space and eventually inserting facilities such as schools and health centres should be encouraged in the bustees. In 1969, a programme was introduced in bustees in which the open space which is 20 per cent in these settlements would be covered by very low cost housing.

Policies of the development plan

In 1969, Colin Rosser expressed, "What Calcutta clearly needed it seemed to us, was not a negative, restricted, regulative land use plan with the heavy commitment of scarce administrative resources that such a system of control requires, but rather a positive development plan in the full sense with the emphasis on a development action directed at priorities of social and economic change, of selected capital investment in the physical infrastructure, of administrative and fiscal reorganization, and of the systematic generation and mobilization of resources.[12]

The policies may be summarized as follows :

(1) Promotion of dynamic growth of the metropolitan economy by increasing production and income with enough employment opportunities and close integration with the economic development of the region.
(2) Development of satisfactory urban environment sustaining with appropriate facilities and services of population of 12.3 million in 1986.
(3) Provision of machinery for planning and its effective implementation.
(4) Strengthening of local self-government and citizen participation in the planning through mobilization of local fiscal resources, civic leadership and voluntary citizen effort.

Areas of intervention

Indian local government is strewn with the failures of development controls. "Red tape-ism" and flouting of the regulations should be controlled to win the failures in planning. Incomes and employment should be increased and should be given proper place in planning. The intricate and interdependent nature of cities should be minimised for success in planning.

The proposed actions fall into the following groups :

(1) The arrest of deterioration and the removal of deficits in areas in CMD should be given priority with particular emphasis on the concentration of Calcutta and its sister city, Howrah.
(2) Partially developed areas of the metropolitan centre should be developed on priority basis.
(3) Some specific new growth areas should also be selected for planning and development.
(4) Aggressive and accelerated development of a new metro-sub-centre is essential which is situated 64 Km north of the existing centre through building and expanding a premature investment in the satellite town and improving its linkage.

(5) The budget should be concentrated on transportation for major investment.

Immediate action programme

The basic Development Plan concludes with a ringing call for a new, greater, and more beautiful Calcutta but with a specific set of recommendations for immediate action. The actions are proposed to produce "not the city beautiful but the city livable". The Immediate Action Programme has three components :

(1) Selection of the subject matter for immediate action (new township development, urban renewal, shelter, utilities, community services and facilities, traffic and transportation, industrial and commercial development) ;

(2) A specific investment programme which should be responsible for the development of each project, and proposed method of financing ;

(3) Proposals for new government agencies should also be framed which initiate and expedite the development plan.

CONCLUSION

Metropolitan planning is very much similar to regional planning as it is, based on the principle on futurology i.e. '*avant garde*' of planning. To check metropolitan growth is the beginning and end of the *communis opinio doctorum*. It is also framed and practised in spatial, temporal and sectoral framework. The spatial perspective of metropolitan planning adopts the micro, the macro and the mega approaches to deal with urban problems in a comprehensive manner. Sectoral and temporal aspects tackle the problems in a more systematic way.

The case studies from the developing countries represent the same problems in every large metropolis. Disintegration of physical and economic planning of the metropolis with the national level is the main problem which hampers the speedy solution of the urban problems. The metropolitan areas are suffering from high density of population, congestion, lack of transportation, housing and environmental pollution. The respective governments are busy to solve these problems on footwar scale. But migration (inward) and high growth of population make all the estimates and projections false. Development of ring towns seems proper solution as it checks migration. It is already in practice in several metropolitan areas.

FOOTNOTES

1. Kuklinski, A.R., *Contributions to Regional Planning and Development*, Development studies, Number 3, I.D.S. Mysore, December, 1971, pp. 47-54.
2. Myrdal, G., "The Necessity and Difficulty of planning the Future Society", *Environment and Change. The Next Fifty Years*, American Institute of Planners.
3. Papaioannou, J.G., "Future Urbanization Patterns in Europe", Paper read at international colloquium on the Mastery of Urban Growth, Mens en Ruimte, V.Z.W. Brussals, December, 1969.
4. Porwit, K., "les Elements Regionaux I ans le Planification a long Terme", *Les problems Spatiaux dans la Planification en Perspective*, Warsaw, 1968.
5. Kuklinski, A., *Goals in Regional Policies and Objectives in Regional Planning*, UNRISD, 68/C. 48, 1, March 1970. See also, A. Kuklinski, *Research on regional Development in the Framework of UNRISD* 697 C. 53, Geneva, August, 1969.
6. Viot, P., "Through Regional Planning Towards Regional Administration", Paper presented at the IULA Seminar on Regional Administration, Prague, 1969.
7. Harmansen, T., "Information Systems for Regional Development Control : Framework for a Research Project," in *Regional Planning : Concepts, Techniques Policies and Case Studies*, Mysore, 1969. See also W. Kawalcec, "Perspective Development of Regional Statistics in the Light of Polish Experience", paper prepared for the Seminar on I.S.R.D., Lund, Sweden, October, 1969. C.L. Leven, "The Role of Regional Data Bank in the Context of a System of National and Local Data Banks". Institute of Urban and Regional Studies, Washington University, I.N.S. 4.
8. More details can be consulted in *Delhi Master Plan*, D.D.A., 1962.
9. Shibli, K., "Metropolitan Planning in Karachi : A Case Study", in *Metropolitan Growth*, Sage, New York, 1974, pp. 109-25.
10. Bhagwati, J., *Economics of Underdeveloped Countries*, Mcgraw Hill, New York, 1970.
11. Row, A.T., "Metropolitan Problems and Prospects : A Study of Calcutta", in *Metropolitan Growth*, Sage, New York, 1974.
12. Rosser, C., *Urbanization in Eastern India : The Planning Process*, Ford Foundation, Mimeo., Calcutta, 1969.

SIXTEEN

F. SPREER AND K. TURKE

LOCAL AND REGIONAL ENERGY SUPPLY CONCEPTS
A Challenge for Urban and Regional Planning

(0) THE heat production and supply sector evidently has the highest conservation potential, which could be mobilized in the coming years. Regional research has done a lot recently to stress the fact that an increased flow between heat produced in industrial (mainly power generation) processes and heat consumed in buildings is a challenging and rewarding task for regional and urban planning in the future. In this context, we have to consider three main aspects :

- The size and location of power stations relative to the settlement structure ;
- Design of optimal technologies for power-heat cogeneration that adapt the local situations ;
- The organizational and political interplay between public authorities and private firms.

(1) The Federal Republic of Germany used to be a "world trade giant", always boasting of a sound surplus in external trade balances. In the last ten years, the situation changed dramatically (cf. Fig. 16.1). While costs for imported mineral oil soared from 10 billion DM in 1970 up to some 65 billion DM in 1980, the balance of payments reached a negative record

Paper Presented at "European Meeting on Applied Urban Research", Essen, Oct. 2-4, 1981.

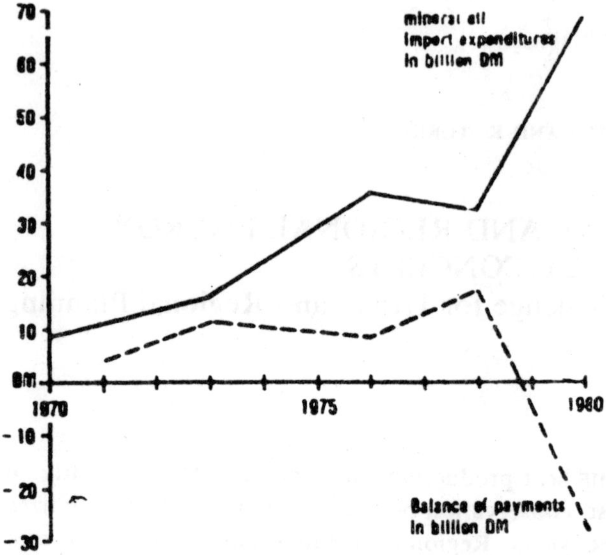

FIG. 16.1 : *External Trade Balances*

of 28 billion DM last year.

In its endeavour to tackle the energy problem, the Federal Government has developed an Energy Program in the course of the 1970s. The current version of that program, updated in 1977, states the following basic objects of the national energy policy :

— limited growth of *energy consumption through economical and efficient use of energy* ;
— reduction of the *proportion of mineral oil* required for energy supplies ;
— preference for use of domestic *hard coal and lignite* ;
— expansion of *nuclear energy* only to an extent absolutely indispensable for electricity supplies ;
— limitation of *importation risks* by a diversification of primary energy resources, international agreements and cooperation ;
— continued *energy research* to mobilize all energy options

possible within the geographical framework of this country).[1]

(2) In order to implement a strategy for energy saving, macro-economic planning and national energy policies have to rely on thoroughly analysed regional and local energy-related conditions and parameters, as well as on local decision making in the public and private sectors that are in line with a "sensible utilization of energy".

For both sides of activities—analyses and implementation—, applied urban and regional research plays an important part. This research functions in two directions :

(1) in analyzing local situations, it can report detailed, implementation oriented information back to decision makers on the State and national levels ,
(2) by disseminating ideas and practical experiences to planners on the local level, it helps pursue national energy policy objectives.
(3) Primary energy input to this country's energy system amounted to roughly 340 million tonnes coal equivalent (tce) in 1977, to 387 million tce in 1978. It is split up into sectors

—mineral oil (53%),
—coal and lignite (27%)
—natural gas (15%),
—nuclear energy (3%).

(4) We must bear in mind the structure of energy use in order to estimate possible primary energy savings and substitution of mineral oil. The following seems to be probable : the sector "households and small consumers" use 45% of the final stage energy in the Federal Republic of Germany (mainly in the heat sector), industries and traffic use 35% and 20% respectively. There is no doubt that households, small consumers and traffic are the most interesting areas in pursuing the two basic objectives of energy policy—saving primary energy

and substitution of mineral oil.

And spatial planning itself is interested in these sectors in particular.

The dimensions of energy to be conserved and saved by urban and regional planning measures is so huge that you could presume that spatial planning arguments would dominate the present discussions on energy policies. But that is not the fact. Instead, almost all of the public attention is absorbed by the electrical power supply and here in particular the nuclear energy issues.

We think that is unappropriate. Electrical power makes up for just 17 per cent of our overall final stage energy demand—heat demand, in comparison, 40 per cent ! In spite of that, heat supply has never been regulated and organized in an appropriate way in this country. The Energy Regulation Act of 1935 is restricted to prescribing a "secure and cheap delivery of electrial power and gas" for the population.

(5) Heat supply, up to now, has been organized exclusively to micro-economic principles of several companies. The main findings of our research are the following :

If additional criteria are combined to form an optimal mix of several heat supply forms, approximately 50 per cent of primary energy for heating can be saved within the next 20 years. These criteria are derived out of an analysis of the given settlement structures, structures and forms of buildings ; they take in consideration optimal combinations of heat distribution and insulation, use of waste and process heat, environmentai heat sources and so on. 50 per cent of energy saving without considerable loss in the standard of heating—that would be more than any other scheme for energy conservation discussed presently can offer. And these calculations are based on quite realistic premises about trends in the future : Growth of housing space per person ; central heating in all buildings ; energy price increased by 200 per cent above the 1978 level.

To differentiate the heat supply systems according to settlement structure, we recorded the present modes of heating in buildings in different types of communities (using data of the Housing Sample 1978).

TABLE 16.1 : **Heating systems in buildings in different types of settlements**

Type of settlement	All housing units (in 1000)	district and block heating (%)	hard coal (%)	mineral oil (%)	gas (%)	electricity (%)
Large cities (>500,000 Inhab.)	886,7	12,6	15,4	38,8	21,5	8,7
Higher level and large middle level central places	7,295,8	8,2	13.2	39,6	25,3	6,0
Middle and lower level central places	9,236,0	2,4	13,1	61,2	11,5	5,7
Towns without central significance	2,352,8	0,6	18,5	66,6	2,5	5,8
All communities	23,771,3	6,1	14,3	49,9	16,9	6,4

Source : Housing Sample 1978, Bundesforschungsanstalt fur Landeskunde und Raumordnung.

Note : (Figures do not add up to 100 per cent because of missing data in some cases.

Table 16.1 shows that pipe-network oriented heat systems are concentrated in agglomerated areas. The energy suppliers of all sectors try to pick out the best parts of the cake ; a reasonable division of labour that leads to modes of heat supply well adapted to the urban and regional situation have been the exception in the past. But just this division of labour is the decisive precondition to pursue the objectives of energy policy. One step in this direction is the joint research program of the Federal Ministries for Technology and for Regional Planning which is known by the name of "Local and Regional Energy Supply Concepts". It is a program to implement existing research findings in a number of specific cases in several cities and regions of this country. The program is organized by the Federal Research Institute for Regional Geography and Regional Planning in cooperation with an institute affiliated to the Federal Ministry of Technology.

(6) The heat production and supply sector evidently has the highest conservation potential which could be mobilized in the coming years. Regional research has done a lot recently to stress the fact that an increased flow between heat produced in industrial (mainly power generation) processes and heat consumed in buildings is a challenging and rewarding task for regional and urban planning in the future).[3]

In this context, we have to consider three main aspects :

— the size and location of power stations relative to the settlement structure ;
— design of optimal technologies for power-heat cogeneration that adapt to local situation ;
— the organizational and political interplay between local and state authorities and public/semipublic/private energy enterprises.

As yet, the power production facilities in this country are concentrated in western parts of Germany, notably in the Ruhr and the Saar areas. This regional bias led to stressed regional disparities : higher environmental damages, a vast unused waste heat potential on one side, higher electrical power prices through transportation losses and other distribution costs on the other (Fig. 16.2).

An increased use of domestic hard coal in standard power plants of 750 MW as well as the construction of large-scale nuclear power plants of 1.300 MW along the Rhine and other rivers might perpetuate the regionally unbalances situation.

(7) There is a number of good reasons why middle and small scale technologies in power-heat coproduction should prevail in the future: security, environment, acceptance. Energy conscious regional planning can add a few substantial arguments :

— As the transport of heat is limited to short distances, power stations should be dimensioned to the heat consumer demands in the area ;
— Middle-sized power-cogeneration is the missing link in

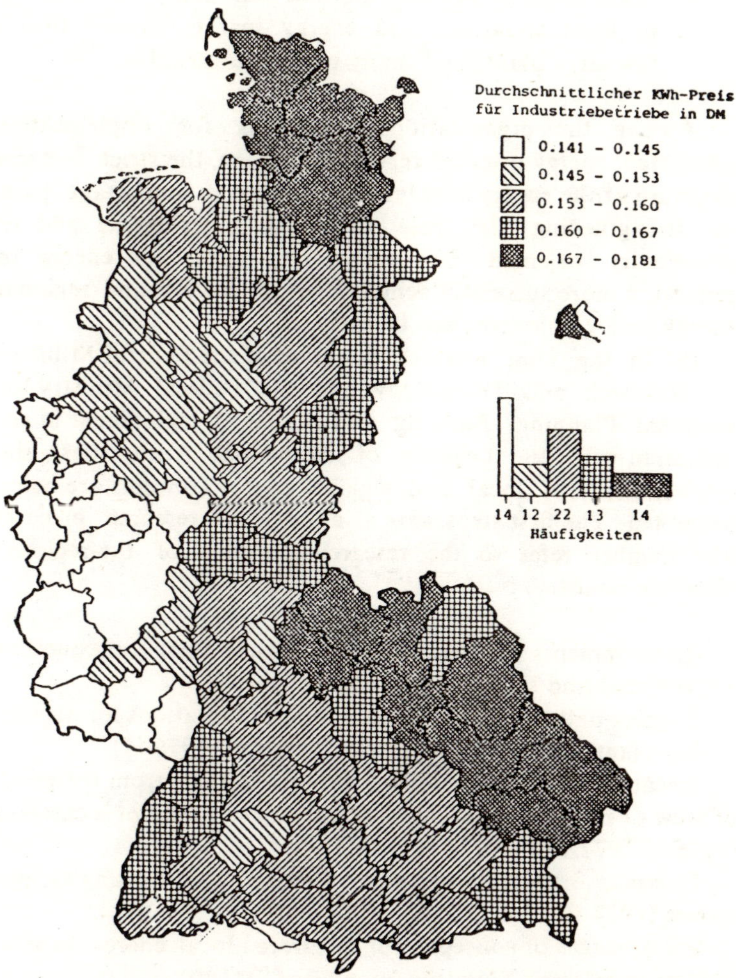

FIG. 16.2 : *Average KWh-Prices for Industrial Use in DM*
(Source : BfLR information system)

a polarized power-heat market : on the one hand extreme centralization of power plants, on the other fragmentation of heat production in individual house (sometimes even room) heating systems ;

— Many small power-heat-stations can adapt more easily to local situations and energy market changes than a few large plants and many private households.

Among the organizational problems for implementing integrated, energy oriented regional planning, the strict sectoral structure of the energy supply industry play an important part. In strengthening the role of local governments and of community oriented companies, responsible for energy in general, a more successful achievement of national and regional energy policy objectives can be expected.

(8) In the joint research program of the Federal Ministry for Research and Technology and the Federal Ministry for Regional Planning, Building and Urban Development in co-operation with the Leagues of Municipal Authorities, the establishment of local and regional energy concepts are being promoted. The first steps were a number of research projects (the number refer to the research program of the regional planning ministry) :

Developments in the energy sector and their consequences for regional and settlement structures (1975.07)

Factors underlying the determination of locations for nuclear energy plants in selected industrial countries (1976.11)

Locational requirements of power stations from the point of view of regional planning (preliminary selection of locations) (1976.17/1977.16)

Interplay between settlement structure and heating supply system (1977.10)

Initial stages of concepts for integrated local energy supply systems in selected European countries (1977.10)

Economical use of energy in the context of local developmnet planning (1978.15)

Economical use of energy in the context of urban renewal (1978.16)

Economical use of energy in the context of new settlement projects (1978.17)

(9) In the next phase—currently under way—a number of model cities and regions will take part in the program with experimental supply concepts to test possibilities and restrictions

for integrated energy planning in a realistic context.

A multidisciplinary steering committee also representing different interest groups will monitor and coordinate the activities. The program is expected to produce the following output :

— comparative description of the low temperature heat market ;
— recommendations for an updated energy program ;
— planning handbook for practical use in local administration ;
— investment plans for municipal energy companies and orientation line for the big energy industries.

(10) Implementing such an ambitious program will ask for huge investments, mainly in district heating networks. In both fields, the federal and state governments have shown their willingness to subsidize such "future oriented investments". According to our planning strategy, district heating systems will be developed "bottom up" : small "island" heating districts will grow together and gradually link to plants at greater distance (Fig. 16.3). Options for such a complex system are restricted to densely populated areas, of course. The "island strategy" is much cheaper than the "strategy of the big pipe".

(11) What will happen to less populated areas, then ? In a regional planning strategy for the whole territory, three main trends might be reasonable to follow :

— substitute gas in agglomerations for district heating (gas companies are still trying to gain territories here) ;
— expand the gas distributing network in less densely populated areas wherever this can be done economically ;
— use alternative energy sources in peripheral rural areas (biogas, wood and straw burners etc.) combined with efforts in insulation.

(12) Energy will be one of the most interesting topics for applied regional research in the coming twenty years.

A great number of detailed questions is still waiting to be answered. These questions are less of a technical nature, although a good deal of knowledge is expected here, too.

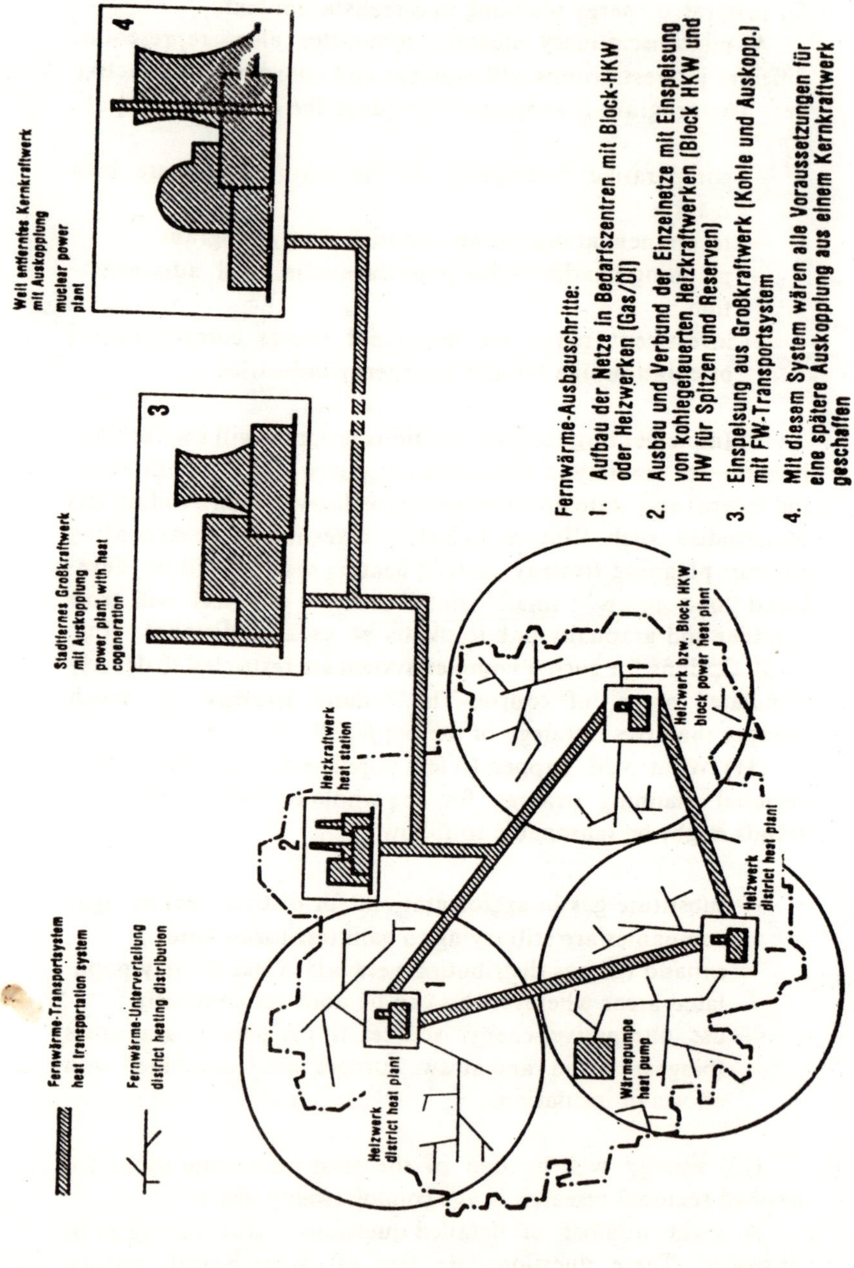
Fernwärme-Transportsystem
heat transportation system
Fernwärme-Unterverteilung
district heating distribution
Helzwerk
district heat plant
Wärmepumpe
heat pump
Helzwerk
district heat plant
Helzkraftwerk
heat station
Helzwerk bzw. Block HKW
block power heat plant
Stadtfernes Großkraftwerk
mit Auskopplung
power plant with heat
cogeneration
Weit entferntes Kernkraftwerk
mit Auskopplung
nuclear power
plant
Fernwärme-Ausbauschritte:
1. Aufbau der Netze in Bedarfszentren mit Block-HKW oder Heizwerken (Gas/Öl)
2. Ausbau und Verbund der Einzelnetze mit Einspeisung von kohlegefeuerten Heizkraftwerken (Block HKW und HW für Spitzen und Reserven)
3. Einspeisung aus Großkraftwerk (Kohle und Auskopp.) mit FW-Transportsystem
4. Mit diesem System wären alle Voraussetzungen für eine spätere Auskopplung aus einem Kernkraftwerk geschaffen

Fig. 16.3

What is more important : the application of a rich variety of technical options to local and regional situations, the promotion of the organizational and political implementation of "energy wise" decisions in face of micro-economic interests, which do not sum up to a macro-economic optimum. It has to be tested.

FOOTNOTES

1. Bundesministerium fur Wirtschaft : Energieprogramm der Bundesregierung. Zweite Fortschreibung Bonn 1977.
2. E.g. U. Roth et al. : Wechselwirkungen zwischen Siedlungsstruktur und Warmever-sorgungssystemen. Bonn 1980=BMBau-Schriftenreihe 06.044.

SEVENTEEN

KARLA M. DYKES

THE LIMITS TO ENERGY-SENSITIVE LAND USE PLANNING, PROBLEMS AND SOLUTIONS IN THE UNITED STATES

Overview

EIGHTY per cent of the world's energy is consumed by thirty per cent of the world's population and the United States with six per cent of the population is consuming over one-third of the world's total energy consumption. In terms of distribution in the United States currently, twenty-eight per cent of the energy utilized is for space or water cooling/heating, an additional twenty-five per cent for transportation, thirty-seven per cent for industry and the remainder for utility consumption (Manohar, 1981).

The existence of shortages of traditional energy sources poses a design challenge to architects, engineers, and social planners. Energy shortages and controls on the use of energy will have an ultimate effect on urban planning inclusive of the location of structures and the uses and functions to be served, the design of industrial, commercial and residential buildings, the transportation systems spatial uses, and overall living patterns of people in urban and rural areas.

The practice of urban planning and architecture has sought to create cities which are more compact, capable of being serviced by public forms of transportation, are efficient in the distribution and consumption of energy, are accessible to open space and recreational opportunities, and are in a sympathetic relationship to the natural environment. While the validity of these arguments was recognized by many, little was done to achieve results. Little was done because without the economic incentive to do sound urban planning the highly speculative land development market, spurred by massive road building produced low density, disorganized, inefficient, energy consuming urban sprawl, in which our society now has a great investment, both socially and economically.

However, a shift in orientation is currently taking place as communities are beginning to take energy planning into their own hands. Realizing the drain on the local economy from dependence on imported fuel, a number of cities and towns are drafting municipal energy programs and providing local incentives for solar use and energy conservation.

Energy sensitive land use controls are in their infancy ; in the United States federal, state and local measures all date from the late 1970's. At the local level numerous communities have effected energy sensitive land use regulations. In many such cases, an energy element or equivalent is first adopted in the local master plan. This is followed by revision of the zoning/subdivision ordinances aimed at realizing the energy conservation goal through such means as increasing allowable density, reducing the minimum size of dwelling units, protecting solar access and activities to be discussed in this paper.

There are many potential benefits to be gained from local community based energy programs. A local government represents a smaller more unified constituency. Having more direct control over zoning and land use, transit planning, and building codes, it has the opportunity to eliminate regulatory barriers, use alternative energy sources in municipal buildings, and develop a community energy plan that involves all elements of the community.

Still, specific linkages of energy and land use remain unstated and unexamined, this is the topic of future research—to build a foundation for the clear explication of the relationships between energy considerations and land use planning. Towards this end, energy and land use may be defined as : land planning requirements to enhance energy supply and reduce energy demand. These requirements recognize both :

(1) the influence and pervasiveness of past development trends, and (2) necessity, in some instances, to redirect past trends (Burchell, Listokin, 1981).

Serious shortages of energy may alter the future development of cities in a much different context and with a much different conformation than that of today. Typical suburban residential densities of four families to an acre may become impossible and some communities may be forced to reject single family dwellings entirely in favor of high density residential development. The American dream of home ownership may be converted to a dream of condominimum ownership.

In transportation planning the most notable shift will be from highway construction to mass transit subsidies in an attempt to lessen reliance on the automobile as the prime means of moving people in their everyday journeys. When private automobiles are used it will be within "ride-sharing" arrangements. The bicycle will gain prominence as a means of transportation as provisions are made for their safe and efficient use through special bike lanes and pathways along with pedestrian movement planning.

However, it should be noted that President Regan is making a public presentation of a cheque to Mayor Koch for federal financing of Western Highway construction in lower Manhattan at a time when mass transit subsidies are sorely needed in New York City. Further, Mayor Koch's experimentation with bike lane island construction—following a visit to China last year—has been torn up and removed the excuse being that the special bike lanes were not being used, despite significant protest by bikers.

Energy and metropolitan development

The literature on energy and metropolitan development primarily is concerned with assessing the evidence supporting or denying the existence of an urban energy advantage. It attempts to place energy implications in perspective with social and other forces influencing the return to the central city. These analyses involve consideration of the changing economic function of the central city ; comparison of the journey to work of urban versus suburban residents and a course the costs and use of the automobile as a means of transportation. Topics also covered involve the urban housing stock, tenant/owner characteristics ; and evaluation of the spatial requirements of never before utilized energy systems.

The question to be asked is whether the location of the most energy efficient geographical locus is located in the older central city (the residual center of mass communication and most fixed rail transit systems) or has it shifted to the inter-states and beltways of metropolitan America served by telecommunications, soft wheel mass transit, fuel efficient automobiles, and diesel trucking ; this again is a topic for further research. There is an increasing trend that identifies the emerging path of energy sensitive land development to be occurring outside of established metropolitan areas. The argument that energy shortages will encourage a return to the central city have been based on the following assumptions :

1. The urban housing stock, because of its density and compactness, consumes less energy (Bacon, 1973).
2. Cities benefit from advantageous transportation and communications characteristics (Fels, Munson, 1975).
3. Cities can readily capitalize from emerging and more efficient energy systems (Knowles, 1974).

Considerable literature argues that cities are more energy efficient for many reasons. Reference is made to cross national studies such as (Doernberg, 1975) comparing the United States and Sweden, indicating that while the two countries shared an equivalent per capita GNP, Sweden consumed sixty per cent less energy per person because of its higher density of uses and more intimate mixing of residential, commercial and industrial activities. Recently urbanists have begun to question whether cities in reality benefit from a current or future energy advantage significant enough to reorient the pattern of land use development.

City housing stock has two energy saving characteristics being high density and often attached, however, this housing is old and often energy inefficient because of masonry or even wood construction, ceiling height, room layout, lack of weatherization, outdated heating/cooling plant and other characteristics. These drawbacks can be corrected by installing insulation, storm windows and new energy systems, but such alterations require financial capacity not available to the urban poor who live in the most energy inefficient dwellings. Therefore, the urban housing stock may prove to be an energy debit rather than credit. These factors stand in contrast to suburban housing stock and the relative affluence of the suburban population.

Further, cities may have an advantage with reference to certain new energy approaches, such as cogeneration but it is problematical how they fare with respect to others. For example, solar energy applications meet with impediments in cities. The height and orientation of city buildings can create access problems. The compactness of cities creates a panalty due to high energy demand relative to the surface space available and needed for solar collection. For example, covering every square yard of Manhattan with solar collectors would supply only about fifteen per cent of the borough's intense demand for energy (Giggons, 1979). In addition, applying solar to cities would require extensive retrofitting on high-rise structures—a more difficult and expensive hurdle than utilizing solar in new, low-rise construction.

Other energy sources are likely not to be appropriate for urban adaption. The combustion of wood and other renewable resources, is precluded by stringent urban air pollution safeguards to prevent further air quality degradation, as with the use of coal ; the by-product being "acid-rain". Wind generated energy is typically unfeasible because of blockage from existing buildings however, it can be utilized along harbors, waterways and lakefronts. The possibility of nuclear plants in or close to urban areas is certainly slim due to the threat to large masses of population.

Optimistically, a University of California study entitled, *An Overview and Critical Evaluation of the Relctionship Between Land Use and Energy Conservation*, is particularly concerned with changes in technology which can alter the land use restraints and opportunities for different energy systems and illustrates that development of low cost insulation materials could allow cogeneration systems to be placed in low density areas. Also, the production of more efficient/smaller solar collectors may lessen the restraints to the utilization of solar in cities (U.S. Federal Energy Administration, 1976).

It should be noted that many white, middle-class families have purposely sought a suburban refuge in order to segregate themselves from the urban poor and minorities. This desire is so strong that it often outweighs any energy argument for an urban return. The factual uncertainty of whether cities have a decided energy advantage makes the energy argument for an urban return even less influential.

Western Europeans have reacted to severe fuel prices by reducing the size and increasing the efficiency of their automobiles and not restricting the outward flow of their settlement patterns. This pattern will likely hold tiue for the United States.

Energy conservation and building codes

The majority of the energy studies of the last decade ; the major studies being the (Ford Foundation Energy Policy Project, 1974 and Harvard Business School, 1974) stress the importance of energy conservation.

The Harvard study concludes that energy conservation measures could reduce energy consumption by thirty to forty per cent and therefore is the primary energy strategy because unlike some energy options, conservation measures do not hinge on a major technological breakthrough and can be utilized without the safety and environmental hazards associated with the use of nuclear and coal. These recommendations for the implementation of energy conservation measures are endorsed by such influential professional and policy studies as the (American Institute of Architects, 1975) *A Nation of Energy Efficient Buildings* and Lovins, 1977 *Soft Energy Paths*).

These studies stress the importance of using land use controls and building codes as mechanisms to foster energy conservation. An energy specialist at the Department of City Planning, San Francisco, California, resigned in 1979 in protest after her three years of work toward a "city energy plan" was totally ignored by department officials. Her proposals included revisions to the City Planning Code to permit the home construction of greenhouses among other energy conserving practices.

Building codes can be worded so negatively that builders may be unable to take advantage of new techniques of energy conservation, such as solar heating, greenhouses and below ground construction.

New York's building codes are a big block to energy economy. The codes are relics of an era when energy and capital were relatively cheap. New York's codes were developed at a time when the city was establishing high performance levels for urban services such as education, parks and libraries. In housing it was hoped that quality could be achieved incrementally as tenements and old wood frame units were replaced by modern structures. As a result a builder in New York City is more constrained by provisions for health, safety, fire prevention, nuisance abatement, and personal security than elsewhere. With all the rules in force it is extraordinarily difficult to reduce net energy consumption per dwelling unit or work place (Darmstadter, 1975). In the book, *Conserving Energy : Prospects and Opportunities in the New York Region*, the author, Darmstadter shows the difficulty suggesting that even under feasible (that is within the law) non crisis conservation measures, overall energy consumption would still increase by two-point-seven per cent per year. Without such measures he projected a three-point-five per cent per year increase in energy use.

Adequate insulation can pay for itself in a reasonable number of years, even in a climate as that of the New York Metropolitan Area (Environmental Law Report, 1975). Commercial buildings through a program of insulation, shading and improvements in the heating, lighting, and cooling systems can save large amounts of energy at costs which appear reasonable (ibid).

The general public must first be convinced of the need to conserve energy and must be provided the information necessary to make efficient choices in their consumption of energy. Much of the energy consumed in the residential and transportation sectors is directly affected by individual consumers' personal habits and behavior. It appears that a conscientious concerted effort by the public to change their energy consumption habits and behavior has not yet taken place in the United States. Convincing the public to make such changes is probably the most important challenge. Also, it must be taken into account that the business of the power utilities is to sell energy at a profit—not to conserve it.

Public initiatives

Public regulations did not specify required energy performance for buildings before the 1973 oil embargo. Presently in New York City one-half of the rent roll of multi-unit structures goes to paying for energy costs. The embargo led the American Society of Heating, Refrigeration and Air Conditioning Engineers, Inc. to formulate in 1974-75 a code (ASHREA Standard 90-75) which establishes energy performance levels for different building components. Through federal prodding, ASHREA Standard 90-75 was incorporated by many states into their official building codes. This was a major breakthrough mandating energy efficiency in new construction. However ASHREA had its shortcomings, being only concerned with building components. The document served as a model for the more comprehensive Building Performance Standards, (BEPS) developed by the federal department of Housing and Urban Development, between 1976-1979. BEPS considers existing energy-saving technology ; familiarity of design professionals with these systems ; the ability of recent construction to satisfy hypothetical levels of energy efficiency ; and varying local factors, such as energy cost/supply and climatic conditions. All new federal buildings will have to conform with the proposed BEPS that builders would have to meet before construction could begin. All new commercial and residential buildings would be subject to these standards and if made mandatory through the imposition of federal sanctions against non-complying states and localities, these standards would become a major part of the codes that every builder must follow.

Many efforts by state and federal authorities to cajole and encourage code conformity and standardization have failed or have been only partially successful. While it is true that the federal government can influence building standards through a combination of moral persuassion and incentives such influence is fairly limited in its total effect on the quality of construction in the United States. The institutional barriers that have so far prohibited the adoption of a model code on a national basis, will continue to create barriers for the implementation of the first major voluntary standards concerning energy utilization in new buildings available for optional acceptance by state and local governments.

Consideration of how BEPS will affect land use patterns and development practices could go either of two directions exampled in two energy-land use scenarios. First, BEPS could markedly affect development patterns by encouraging more energy efficient multifamily construction and changing traditional building orientation to maximize solar potential. Or, BEPS would minimally affect land use patterns because its energy standards could be satisfied by changes to the building envelope itself, by such means as improving insulation in ceilings/ walls and utilizing insulated glazing (Thompson, 1981).

Determinants of energy consumption and land use studies

The emergence of new public regulations has sparked studies that consider and compare approaches to lessening energy consumption which are most comprehensive and innovative. The literature concerning land use and energy consumption has dealt with the related subjects of the ability of land use to foster energy conservation and land use determinants of energy consumption.

The methodological approach of early energy consumption/ land use studies is typified by the widely cited, *The Costs of Sprawl* (Real Estate Research Corporation, 1974). This investigation considered different categories of communities : 1) low-density sprawl, consisting of conventional single family development ; 2) high-density-planned, composed of multi family/ attached units with only ten per cent single family ; and 3) a combination mix blending different categories of units and development modes, both conventional and planned. The study concluded that the high-density-planned scenario offered an energy savings of over forty per cent as compared to the low-density sprawl case. The study's methodology has been questioned by urbanists who feel that many of its assumptions and conclusions were oversimplified and as such the development scenarios do not adequately mirror real world development. Despite these criticisms, *The Costs of Sprawl* remains the most widely cited analysis of its kind.

Energy consumption can be regulated by land use controls placed on site design. Orienting structures on a southwest to southeast axis maximizes the polential for active and passive solar heating. Emphasis on linear building arrangement reduces energy losses by providing protection from the wind ; taking advantage of natural and added vegetation shelterbelts which reduce energy waste. Also, energy consumption for cooling purposes is lessend by vegetation shelterbelts and by staggered building arrangements which accentuate the cooling provided by breezes. Extensive research into the potential and implementation of energy-sensitive site design control are outlined by (Olgay, 1963) *Design with Climate.* The (Argonne National Laboratory, 1979) SAND (Site and Neighborhood Design) study has documented actual energy savings in five case communities representing a variety of climatic settings, development programs, and schedules.

These studies reflect a diversity of approaches to energy consumption at the scale of site and neighborhood design. SAND has established that substantial energy savings can be achieved by modifying site designs during the development process. Estimated annual energy consumption reductions were as high as sixty per cent in some communities studied.

The most comprehensive study has been undertaken by researches Erley and Mosena of the American Planning Association, providing a national review of communities enacting energy-sensitive measures. They surveyed some 1,500 local, regional, and state planning agencies. From this data they identified thirteen jurisdictions with particularly innovative energy conserving development regulations which they examined at length, addressed in their book, *Energy Conserving Development Regulations : Current Practice.* Erley and Mosena classify the land use statutes they found into the following four groups :

(1) The most common category moderates heating/cooling needs through such means as : encouraging/allowing

energy efficient dwellings, including multifamily and earth sheltered housing units ; and modifying subdivision requirements to allow for solar energy use and natural cooling/wind protection.

(2) The second group reduces transportation related energy consumption by encouraging compact development and a mixture of residential/nonresidential uses features permitting fewer trips in general and facilitating mass transit in particular.

(3) The third category lowers energy consumption in construction materials and processes. Permitting higher density development with attendant reduction in the length of necessary utility and other infrastructure requirements is a method of accomplishing this.

(4) The final land use group turns from lowering energy demand—the focus of the first three groups—to encouraging alternative energy sources and systems, such as solar and wind generating energy and district heating cogeneration.

In their study analysis the authors (Erley, Mosena, 1980) go beyond innovative energy land use measures and review their planning implications discussing the "trade-offs" of energy regulatory scope and acceptance. Finding and suggesting that comprehensive controls covering all aspects of development offer the greatest potential for energy saving, but in so drastically challenging statue quo, are least likely to be accepted. More palatable, politically, though offering less drastic energy benefits, are incremental/partial energy conserving regulations. The authors stress the importance of planners in the process to develop a constituency for energy-sensitive land use controls through education, experimentation and the utilization of different approaches.

End, energy in context

Those of us who deal with legal and technological devices—seeking to anticipate, to think beyond our time—must keep in mind that the end result of our efforts is to advance the human qualities and enjoyment that make life worthwhile. While the excessive energy consumption induced by sprawl is significant, it is only one of the many pressing concerns to the well-being of metropolitan areas, additionally on the list and interactive, are the adjoining crisis of central city decline ; environmental degradation ; fiscal insolvency ; excessive agricultural land consumption ; and increasing housing costs. Therefore, land use planning must seek resolution in all these problem areas along with the application of proven energy-sensitive land use planning techniques.

REFERENCES

American Institute of Architects (1975). *A Nation of Energy Efficient Buildings by 1990*. American Institute of Architects. Washington, D.C.

American Institute of Architects Journal (November, 1975). "Redesigning an Entire Town (and Its Lifestyle) For Energy Conservation". American Institute of Architects. Washington, D.C.

Argonne National Laboratory (1979). *Community Systems Energy Saving Programs for Committees*. Argonne, Illinois.

Bacon, Edmund (1973) "Energy : shaper of future living patterns." *AIA Journal*, December 1973.

Beuscher, et al (1976). *Land Use second edition*. West Publishing.

Booz, Allen, and Hamilton, Inc. (1976). *Interaction of Land Use Patterns and Residential Energy Conservation*. Federal Energy Administration, Washington, D.C.

Burchell, Robert and David Listokin, editors (1981). *Energy and Land Use*. Rutgers University Center for Urban Policy Research. New Brunswick, New Jersey.

Darmstadter, Joel (1975). *Conserving Energy, Prospects and Opportunities in the New York Metropolitan Region*. Johns Hopkins. Baltimore, Maryland

Erley, Duncan, and David Mosena (1980). *Energy-Conserving Development Regulations Current Practice*. American Planning Association. Chicago, Illinois.

Fels, Margret F., and Michael J. Munson (1975). "Energy thrift in Urban transportation : Options for the future". In Robert H. Williams, ed. *The Energy Conservation Papers*. Ballinger. Cambridge, Massachusetts.

Ford Foundation Energy Policy Project (1974). *A Time to Choose : America's Energy Future*. Ballinger. Cambridge, Massachusetts.

Giggons, John (1979). Director, Office of Technology Assessment, Testimony before the Subcommittee on the City of the House Committee on Banking, Finance and Urban Affairs. October 16, 1979. Washington, D.C.

Hagman, Donald (1974). "Planning Underground Uses", in, *Legal,*

Economic and Energy Considerations in the Use of Underground Space. National Academy of Sciences, Washington, D.C.

Harvard Business School (1974). *Report of the Energy Project.* Harvard, Cambridge, Massachusetts.

Harwood, Corbin Crews (1977). *Using Land to Save Energy.* Ballinger. Cambridge.

Keith-Jones, Gail (1981) Director of Planning, Ecologic. 150 Columbia Heights, Brooklyn, New York 11201.

Knowles, Ralph (1974). *Energy and Form : An Ecological Approach to Urban Growth.* MIT Press. Cambridge, Massachusetts.

Little, Arthur (1975). *Residential and Commercial Energy Use Patterns 1974-1990.* Federal Energy Administration. Washington, D.C.

Lovins, Amory B. (1977). *Soft Energy Paths : Towards a Durable Peace.* Friends of the Earth. San Francisco, California.

Olgay, Victor (1963). *Design with Climate : Bioclimatic Approach to Architectural Regulation.* Princeton, University Press, New Jersey.

Real Estate Research Corporation (1974). *The Costs of Sprawl : Environmental and Economic Costs of Alternative Residential Development Patterns at the Urban Fringe.* U.S.G.P.O. Washington, D.C.

Tarlock, Dan (1974). "Legal Aspects of Use of the Underground," in *Legal, Economic and Energy Considerations* (ibid.).

Thompson, Grant (1981). "Federal Building Energy Performance Standards", in Burchell, Listokin eds. (ibid).

United States Federal Energy Administration (1976). *An Overview and Critical Evaluation of the Relationship between Land Use and Energy Conservation, 3 vols.* Washington, D.C.

Wright, Robert (1974). "Development of Policy for Use of Airspace", in *Legal, Economic and Energy Considerations* (ibid).

EIGHTEEN

HELMUT GUTTLER

THE APPLICATION OF LAND POLICY INSTRUMENTS IN URAN PLANNING

Definition of land policy

LAND policy may be defined as the sum of all political measures which influence land ownership, land use and the distribution of the revenue on land.

Major criteria governing land policy measures in the Federal Republic of Germany include

- the supply aspect, i.e. the aim to make available cheap building land for the housing requirements of large sections of the population,
- the allocation aspect, i.e. the utilization of land policy and fiscal instruments with the aim of manipulating land ownership according to the provisions made in local and sectoral plans in such a way that plots are used for their designated purpose and making available undeveloped building land for urban land use through the owner or third parties.

The urban planning system according to the federal building law

Development planning in the Federal Republic of Germany is structured in two levels :

Paper Presented at the "European Meeting on Applied Urban Research" Essen, Oct. 2-4, 1981.

The *land use plan* (*Flächennutzungsplan*) gives a general framework óf land use on the basic of urban development objectives for a medium or long-term period. The land use plan is not immediately legally binding for citizens, but it binds communes as well as sectoral planning agencies which have not objected to the plan during the public hearing, although amendments or the scrapping of the plan are still possible.

The obligation of the communes to provide a development plan as a first step towards the land use plan was originally envisaged, but not included in the Federal Building law. Nevertheless all existing communal development plans must be considered in the setting up of the overall land use plans, as long as they are relevant to the planning process.

The second step is the *local plan* (*Bebauungsplan*).

It is developed out of the land use plan and gives a detailed determination of urban land use which is binding for everyone.

Changes in land use and building projects which require planning permission are only allowed in the area covered by a local plan, if they do not contradict the plan or if they are subject to special regulations or exemptions.

The local plan is the basis of a number of supporting and implementing measures (see No. 3b). In order to serve these functions it must at least contain provisions about type and extent of building use, available building space and local traffic space.

The *Building Use Regulation* (*Baunutvungsverordnung*) contains regulations about type and extent of building use.

The planning regulations are of importance for the building market and land value assessment insofar as they determine the development condition (the quality) of the plots in the planning area according to location, type and potential use.

The local plan is passed by the communes and requires, like the land use plan, the approval of superior authorities.

Land use plans and local plans are produced by communes at their own responsibility, as soon and as far as it is necessary for urban development and planning. This principle of communal planning sovereignty implies *planning duties* as well as *planning rights*.

The *Federal Urban Renewal Act* contains a special set of conditions under which there is an obligation to set up lccal plans in formally determined improvement and development areas.

Communal planning sovereignty is limited by the Federal Building Law insofar as land use plans and local plans have to consider the aims of regional planning. During the planning approval procedure superior authorities investigate whether, and how far, these aims have been net.

A further basic legal principle is the "*balancing order*" (*Abwagung‹gebot*), which incorporates three main aspects : Public considerations are to be balanced with each other, private considerations are to be balanced with public ones, and private considerations are to be balanced with each other in the setting up of land use plans and local plans. These considerations are reviewed by the chambers for building land matters in the public courts.

Public authorities and institutions (mentioned individually in the administrative orders of the federal states) are to be involved as early as possible in the setting up of land use plans and local plans in order to be able to inform the communes on planning measures at hand.

Citizens are to be involved in plan development before the plan is made available for public inspection, so that they can voice doubts and make suggestions. Aims and objectives must be made clear to citizens in the process (*participation*). The aim is to promote public involvement as a counterbalance to the increased planning competence of the communes created in the amendments to the Federal Building Law. The public is to take an active part in the decision-making process, and is to be protected from possible negative planning consequences. Participation is to make the planning process more transparent and to promote public involvement in the development and planning of the commune.

Finally mention should be given to the *principles for social measures*, which are to be proposed where the local plan can be expected to have adverse effects for the people who live and work in the area. A *social plan* is to present the results of the discussion with the persons involved, and should set forth new planning proposals.

Instruments of land policy

Fiscal instruments to improve the supply of building land

The use of fiscal measures to exert pressure on the sale and/or the utilization of land to provide for a more efficient use of space on building plots has again become a subject of discussion in recent years, however, no corresponding legislation has been passed.

Merely, the first version of the Federal Building Law from June 23rd, 1960 included a so-called "*Building Land Tax*" (*Baulandsteuer*), which provided for an increase in the tax index on derelict building land to 20 % and for a further increase to 30% after 4 years. Thus a higher rate had to be paid on these building sites, which encourage the owners either to build on them or to sell them. However, the Building Land Tax remained in force for only 2, 5 years and was lifted again for allegedly constitutional reasons with effect from January 1st, 1963.

The *property tax* in force in the Federal Republic at present is very law. It is calculated by multiplying the standard site by a tax rate (3, 5 %) set by the government and by a collection rate set by the local authorities (eg. 300 %) ; (for emample : Standard site value=10.000 DM ; Property tax=10.000 DM× 3, 5/1000×300/100=105.—DM p.a.).

Since the standard site values relate to a period back in time (the last general assessment took place in 1979 and established an increase in values of about 40 per cent against January 1st, 1964) they do not correspond to present market values, but amount to about 10-20 per cent of the market value.

Hence the property tax is no longer a significant burden to owners of building land, and therefore useless as a instrument of land policy. There have been discussions on whether the tax burden on plots should be increased by means of a *tax on land value* (*Bodenwertsteuer*) or a *property increment tax* (*Wertzuwahssteuer*) for policy reasons (for example : Market value of site= 100.000 DM ; Tax on land value=100.000 DM×30/1000× 300/100=9.000 DM p.a.).

Legal land policy instruments

Legal land policy instruments are of greater importance than fiscal instruments. There are standardised regulations for town planning in the *Federal Building Law* (*Bundesbaugesetz*) and the *Federal Urban Renewal Act* (*Städtebauforderungsgesetz*).

Legal land policy instruments aid the establishment of plans and the implementation of planning.

In most cases the existence or at least the setting up of a local plan setting forth minimum requirements (type and extend of building use, delimination of developable sites and traffic space) is the precondition for the use of urban land policy instruments.

Land policy aims of the Federal Building Law

The land policy aims of the *Federal Building Law* are based on the principle that measures for the creation of the functioning building land market should not run contrary to the principle order of the market economy. Nevertheless regulations are necessary to ensure that additional building land can be made available, reallocated and fitted with infrastructure if required, and that building land is mobilised through an increase in the pressure of demand for building land.

The primary aim of land policy from urban point of view is the improvement of building land supply, a space-efficient allocation of land, and the ensurement of rapid urban land development.

Concepts include amongst other things an extension of building land reallocation with increased provision of land for third parties willing to build, communal development measures in areas of particular importance for urban development, fewer restrictions in the use of building enforcements to close building gaps, and suggestions for space-efficient building and service provision.

Legal instrument in the establishment of land policy

The following measures may be used in the establishment development plans :

—*The supervision of land transactions* during the setting up of local plans by means of a ban on changes in ownership, the deferment of requests for building permission in special cases and permission to subdivide plots. The aim is to prevent private projects which would run contrary to the aims of the local plans (such as the erection of buildings or the subdivisions of plots) over a given period of time (up to 4 years).

—*Rights of preemption* for local authorities. The aim is to allocate sites for specific purpose if the owner is not able or willing to comply with local authority aims, to ensure that planning objectives are met, or to acquire land for exchange purpose, for instance to facilitate land reform measures. The right of preemption is "price-limited", *i.e.*. sales are being conducted at present market value.

Legal instruments in the implementation of Land Policy

Instruments for the implementation of land policy in urban planning include the following :

(1) the *reparcelling* (reallocation of building plots),
(2) the *boundary regulation* and
(3) the *expropriation.*

It is general practice that the purchase or the reparcelling or site boundary regulation are to be applied first as the "softer options" in the implementation of the local plans before expropriation can be enforced.

1. *Expropriation* may be applied only in a limited number of cases, when

—the aim of expropriation cannot be attained by another, more acceptable option, e.g. the exchange or purchase of sites or a reparcelling,
—if it is in the interest of public welfare and
—if the expropriated site can be allocated to its projected purpose in an acceptable time period.

2. The *reparcelling* is a land-exchange procedure for the development of plots. It is regulated by law, and has the aim of producing plots which comply in location, shape and size as well as their use (building and otherwise) with the objectives of the local plan. Principles governing the reparcelling include the principle of expedience, the principle of property, the principle of relativity (all owners have to provide an equal share of land

for communal space) and the principle of equivalent value (no land owner should gains or losses in wealth as a result of the reparcelling).

The increases in land value resulting from the development of the sites are distributed in the form of money or in land area (where the gain in space may reach a maximum of 30 per cent of the land originally contributed).

The third amendment to the Federal Building Law (1981) expands on this : According to the preliminary version of the law the gains in land value resulting from an "*extended reparcelling*" in areas with an urgent need for building plots should accrue to third parties who are willing to build in the area. Since extended reparcellings are implemented above all in local plans which aim at "spatially efficient building".

3. In areas covered by a local plan, in areas subject to reparcelling and in some cases also in built-up areas not covered by local plans the following policy instruments may be enforced :

—*building orders*, which commit the land owner to build on his site within an appropriate time limit,

—*utilization orders*, which commit the owner to use his site in accordance with the provisions made in the local plan, for instanee to prevent the existence of empty housing,

—*maintenance orders*, amongst the existence of empty housing, displacement of low-income groups from their housing by discouraging expensive modernizations with resulting rent increases,

—*modernization orders* to repair and modernize buildings.

However the application of these orders by local authorities is often difficult, as land owners may require the communes to purchase their plots at market value, because they are unable to comply with the orders for objective or subjective economic reasons. However the communes are often unable to purchase for lack of finance.

4. Apart from the reparcelling it is possible to use *site boundary regulations* in some instances, where only the site

boundaries need to be altered in order to enable development to go ahead.

Land supply policy (land reserve economy)

Finally one can remark that the land policy instruments in the implementation of planning can be supported by the *Public Land Supply Economy* through the sale of building land at low prices by communes or central government.

Thus a so-called "Site-Cost Reduction Law" is in operation in the Federal Republic, according to which land owned by the government may be sold below market price to encourage the development of private property among large sectors of the population.

An important feature of land supply policy is that sites are being bought in time of law demand and sold when the demand is high. There is a considerable number of local authorities in the Federal Republic which operate long term land supply policies and thus effectively support housing programmes.

Further development of real estate legislation and land policy instruments

A number of points on this aspect should be made, as far they have been or are being discussed :

(1) In the 1970's the *planning value adjustments* (compensation and betterment levies) played an important role in the first amendment of the Federal Building Law (1976). The planning value adjustments were not included in the *Federal Building Law*, but in the *Federal Urban Renewal Act*: Increases in value resulting from the anticipation of urban renewal or development programmes in formally established urban renewal areas or development areas are not longer considered in the purchase of sites. Furthermore the owner must pay a betterment levy for increases in site value after renewal has been accomplished.

(2) Apart from the tax on building land mentioned above many suggestions have been, and continue to be, made for a *land value tax* (Bodenwertsteuer), or a *property increment tax* (Wertzuwachssteuer) on building land.

(3) Further mention should be given to the idea that the contributions payable towards the costs of the development of local communications and open space should be replaced by a *general levy on infrastructure* (either as a single or as a continual payment).

These three major suggestions for the further development of real estate legislation and land policy instruments, which have led to controversies also for distributive considerations, are to conclude the brief contribution to land use policies.

REFERENCES

Bundesbaugesetz (*BBauG*) of June 23rd, 1960 (BGBl. I, p. 341) in the version of the public announcement from August 18th, 1976 (BGBl. I., p. 2256).

Stadtebauforderungsgesetz (*StBauFG*) in the version of the public announcement from August 8th, 1976 (BGBl. I, p. 2310).

INDEX